terra madre

terra madre

**FORGING A NEW GLOBAL NETWORK OF
SUSTAINABLE FOOD COMMUNITIES**

Carlo Petrini
Founder and President of Slow Food

Foreword by Alice Waters

CHELSEA GREEN PUBLISHING
WHITE RIVER JUNCTION, VERMONT

Copyright © 2009 Slow Food® Editore S.r.l.
Via della Mendicità Istruita, 14-45
12042 Bra (Cn), Italia
www.slowfood.it

Copyright © 2009 Giunti Editore S.p.A. Firenze - Italia
www.giunti.it

Author: Carlo Petrini

Original Italian edition published in 2009 by Giunti Editore S.p.a., Milan, and Slow Food® Editore Srl as *Terra Madre: Come non farci mangiare dal cibo*.

All rights reserved. No part of this book may be transmitted or reproduced in any form by any means without permission in writing from the publisher.

Translated by John Irving

Project Manager: Patricia Stone
Developmental Editor: Benjamin Watson
Copy Editor: Cannon Labrie
Proofreader: Helen Walden
Designer: Peter Holm, Sterling Hill Productions

Printed in the United States of America
North American edition first printing February, 2010

Our Commitment to Green Publishing
Chelsea Green sees publishing as a tool for cultural change and ecological stewardship. We strive to align our book manufacturing practices with our editorial mission and to reduce the impact of our business enterprise on the environment. We print our books and catalogs on chlorine-free recycled paper, using vegetable-based inks whenever possible. This book may cost slightly more because we use recycled paper, and we hope you'll agree that it's worth it. Chelsea Green is a member of the Green Press Initiative (www.greenpressinitiative.org), a nonprofit coalition of publishers, manufacturers, and authors working to protect the world's endangered forests and conserve natural resources.
  Terra Madre was printed on Natures Book Natural, a 30-percent postconsumer recycled paper supplied by Thomson-Shore.

Library of Congress Cataloging-in-Publication Data
Petrini, Carlo, 1949-
  Terra Madre : forging a new global network of sustainable food communities / Carlo Petrini.
    p. cm.
  ISBN 978-1-60358-263-6
  1. Slow food movement. 2. Food supply. 3. Local foods. 4. Food--Quality. I. Title.

HD9000.6.P49 2010
338.1'9--dc22

2010001012

Chelsea Green Publishing Company
Post Office Box 428
White River Junction, VT 05001
(802) 295-6300
www.chelseagreen.com

contents

Foreword by Alice Waters, vii
Acknowledgments, ix
Introduction, xi

1. All About Terra Madre, 1
    How the Idea of a Terra Madre Network Was Born
    Terra Madre: Poetry and Politics
    The Values of Terra Madre

2. Food Communities, 27
    Communities Worldwide
    Town or Country?
    Parishes
    Back to Basics: "Man Eats Food"

3. Food: Dr. Jekyll and Mr. Hyde?, 45
    The Right to Pleasure
    Tradition and Innovation
    True Unsustainability: The Problem of Paradox
    The Foundations of a New Gastronomy

4. The Value and Price of Food, 61
    Food Is Eating the Environment
    Food Is Eating Farmers
    Donation as Waste Prevention
    Speed, Abundance, and Induced Needs
    Eating Well Isn't Expensive
    Overcoming Uncertainty

5. Food Sovereignty, 83
    (Re)conquering Civilization
    Sovereigns of Production
    Sovereigns of Sustainability
    Subsistence
    Diversity
    Synergy
    Recycling and Reuse
    Decentralization
    Biodiversity and Identity
    Voluntary and Free Trade
    Seed Patents, Monopolies, and Privatization
    Existential Sovereignty
    Sovereignty of Knowledge Systems
    Economic Sovereignty
    New Rights and Participatory Democracy
    Deindustrializing Food
6. Local Economy, Natural Economy, 117
    Home Economy
    The Advantages of the Local Economy
    The Enjoyment of Life
7. The Future of Terra Madre, 149

    A Letter from Enzo Bianchi, Prior of Bose, 153

foreword

Carlo Petrini is, above all, a listener, a collector of stories. As the founder of Slow Food he has spent a quarter of a century listening carefully to the world's humblest and yet proudest people, those whose voices have become faint—its small-scale farmers and food producers. Terra Madre as a global network (or, more accurately, as a series of interrelated networks of local food communities) represents the collective wisdom of farmers, fishermen, nomadic shepherds, artisans, and cooks from 153 countries. Carlo is the prime mover of this vital movement and he serves as the conduit for this shared wisdom. In this book he forcefully makes the case for the urgently needed paradigm shift in our relationship to food. However, his message is far from apocalyptic; instead it is one full of hope, a joyful celebration of the powerful cultural and natural diversity of the planet.

Terra Madre returns food to the center of our culture. It reclaims quality as a democratic right and dismisses the notion that food is merely fuel without pleasure or ritual.

Carlo has been a constant source of inspiration over the twenty years I have known him. We do not speak the same language, but we understand each other perfectly. He is able to communicate an unspoken language. It is a language of shared experience, responsibility, and *joie de vivre*.

When the true caretakers of Mother Earth (Terra Madre) meet in Turin every two years, that same sense is everywhere. Language barriers are easily overcome, cultural differences are celebrated, and long-held secrets of land and sea are shared. Thousands of farmers, fishermen, academics, chefs, students, activists, and artisans from every corner of the planet, many of whom have left their home for the first time, for two days come together to build connections, share knowledge, and contribute to an ever stronger network.

This book is the logical follow up to *Slow Food Nation,* in which Carlo articulates the vital importance of our food choices within the kind of holistic, uncompromising context that I—and so many others—have been searching for. With *Terra Madre,* Carlo gives us all a clear blueprint for a sustainable future.

With hopefulness,
ALICE WATERS
Berkeley, California

acknowledgments

In recent years, my friend and assistant Carlo Bogliotti has helped me develop many of the ideas and achievements described in this book. I am deeply grateful to him for his precious collaboration.

I am likewise indebted to another friend, John Irving, for his work on the adaptation of the book into English.

Special thanks, finally, are due to Stefano Scarafia and Paolo Casalis, who produced and directed the *Terra Madre* DVD, which documents and reveals the faces and the spirit of the Terra Madre event. Their film is the perfect visual complement to my words in this book.

introduction

The "Terra Madre—World Meeting of Food Communities" event was organized for the third time in 2008. The opening ceremony was held on October 23 of that year at Turin's Palaolimpico arena, one of the venues for the 2006 Winter Olympics. It was attended by 7,000 delegates from 153 countries, representing about 1,600 "food communities": farmers, fishermen, artisans, and nomad shepherds young and old, plus traditional musicians, cooks, chefs, and academics. They had flocked from all over the world to listen to the inaugural speeches, and then, over the following three days, to take part in workshops, meetings, forums, talks, and celebrations.

What follows is the transcript of my speech at the opening ceremony to that wonderful audience—a mass of folk teeming with diversity, pride, and enthusiasm. It will, I hope, place in historical context the trains of thought and the ideas I develop in this book, helping the reader to grasp the historical significance and scope of Terra Madre, a network that is rapidly becoming one of the largest at the service of planet Earth.

It's a great pleasure to see you all here again at this extraordinary get-together of ours. To think that four years have gone by since we first staged this wonderful event. This third meeting signals the growth of the Terra Madre network, though it would actually be more proper to speak about Terra Madre *networks*, seeing that farmers, fishermen, and nomad shepherds were joined two years ago by cooks and academics from all over the world, and this year we are also welcoming natural-fiber producers and musicians, all of them peasants and farmers. They bring with them the music of Terra Madre to prove that agriculture isn't just another sector of industry like iron and steel, say, but something much more complex than that. In reality, it is the fruit of a holistic vision that takes in "sacredness" of food, respect for the environment, sociality, conviviality, and culture.

This year Terra Madre welcomes into its midst over 3,000 young people—all students, farmers, and cooks—here as a symbol of the future and, through the passion they intend to ignite within farming communities, to give hope to Mother Earth.

Over the last few years, we have realized one thing: namely, that the seed we have scattered round the world is bearing fruit and growing. Yes, Terra Madre is developing all the time, and now embraces 153 countries from all over the world, supporting thousands of school gardens, farmers' markets, and a new alliance with consumers, or "coproducers," as we intend to call them from now on. But that's not all. Over the last few years you have also organized a number of Terra Madre events in your own countries—in Brazil, in Ireland, in the Netherlands. And now Terra Madre communities have developed in at least 30 countries. The network and networks are getting larger and gaining in strength.

We have to be aware that what has happened and is happening now, in 2008, is the stuff history is made of. Back in 2004, few of us imagined that things would evolve as swiftly as they have. Nor could we have imagined the havoc that is being wrought throughout the world. Or the economic crisis that is strangling social relations, people's daily lives, and politics.

We'll remember 2008, most of all, because, in the first part of the year, we received definite proof that multilateralism doesn't work. In January, in fact, a FAO [UN Food and Agriculture Organization] summit in Rome acknowledged that its goal of halving the number of malnourished and starving people in the world was not and is not achievable. On the contrary, the number of those suffering from hunger and malnutrition is about to top the one billion mark: that means one person on the planet out of every six. This is an epoch-making debacle, and it's happening because all the world's wealthy countries put together have failed to keep their promise to pay an annual contribution of $30 billion. To think that, over the last fortnight alone, they've managed to put together $2,000 billion [$2 trillion] to bail out banks in distress on account of rogue finance! They couldn't come up with 30 billion a year, but they forked out *2,000 billion* in a fortnight. In the face of all this, protest isn't enough. What we have to do is stand up and manifest our outrage, once and for all.

If the FAO summit failed, so did that of the World Trade Organization. It's no coincidence that the dispute there was mainly over duties on foodstuffs. The world's powerful just don't seem capable of coming to an agreement on the food question. They agree on other things, and they stage emergency meetings to save the ailing economy. Yet, at the same time, they fail to do anything constructive about food. In mid 2008, after speculating in poor people's homes, energy, and oil, dealers in what many have defined as "creative finance"—but what I think of as "rogue finance"—ultimately decided to speculate in food and foodstuffs. In next to no time, the price of rice, grain, and corn rose fivefold, triggering repercussions all over the planet.

Here in Italy, we spend 15 percent of our income on food, but in many countries they spend 50, 60, and, in some cases, 80 percent. The impact of this can be catastrophic: in fact the number of malnourished people has increased by a hundred million in one year alone, and food riots have flared up in fifty countries. Now the bubble has burst, and speculation in real estate, oil, and food has fallen flat. The crisis is now systemic, and the whole of the economy is suffering a debacle that will not soon be forgotten. People who think this is a passing crisis are wrong: it's a profound one and is sure to last for many years to come.

Today many of us have mixed feelings. The first is one of great concern: for our future, for our daily lives, for our homes—and for the dignity of the world's poorest peoples. From a different point of view, we also feel a tingling sense of freedom. The time had come for the hype to stop. The time had come to put a stop to the disgraceful spectacle of people unashamedly getting rich quick and looking down on the humble labor of others as marginal and secondary. The time had come for the speculative bubble to burst.

We have to be careful, though. Many of the analyses that are currently circulating are way off the mark; we need to figure things out and analyze the situation thoroughly. In my view, an analyst who thinks the market economy is finished is wrong. It's just not true—anything but. Here's hoping the market economy manages to regenerate itself virtuously, keeping its feet on the ground and connecting more with the rural economy.

People who think this situation is the handiwork of a bunch of con-men alone are also wrong. No, the gangrene has spread everywhere—into politics, into life, into the minds of many people. So never before has it been so imperative to reflect, to reason, and work things out—to try to develop a deep understanding of things. We can't be like the hard of hearing, who sometimes burst out laughing twice: the first time because they see people around them laughing, the second because somebody explains to them why they were laughing in the first place! Before we start laughing, before we express our positions, we need to understand exactly how the world will work with its new economy.

Of one thing I'm certain: the ongoing crisis will lead to greater respect for the rural economy. There'll be more time for agriculture and more time for the real economy, which is the economy that counts. An economy with its feet on the ground and calloused hands—the one you all represent. We'll start appreciating manual labor and the know-how it requires once more; and craftsmanship and small-scale manufacturing work, too. We'll also start caring about the people who work the land again, as well as taking an interest in new technologies at the service of sustainability, the environment, and the quality of life.

This meeting of ours embraces all these topics. Setting out from food, we also tie in agriculture, climate change, and new forms of sustainable clean energy. For years, natural subsistence economies were written off as marginal. For years, this kind of economy was derided. But this is precisely the kind of economy that will save the planet from the crazy market economy and rogue finance. You can rest assured of that.

In the near future, politics and economics will grow aware of the vital relationship between food, agriculture, climate change, and health care, of the landscape and of the beauty of ecosystems—all interconnected problems. They will also grow aware of the mistake they made with intensive agriculture, which has driven millions of the women who were the lifeblood of subsistence farming from the countryside. Earth is a Mother above all because she is worked by the hands of women. To chase women out of the process of agricultural production is a crime.

Consumers will help us in our task. Many are worried about the consumption crisis, but I believe that consumers are simply getting ready to make important choices. They will start looking for local, healthy, fresh, seasonal food; it will be a virtuous, large-scale process. If you get ready too, the food you produce will bring you rich rewards. But we have to make sure that no one takes over your labor to transform quality food into a luxury. Quality is a right for everyone. Beware of people telling you to produce organically, to produce quality food to conquer the markets of the rich. You have to produce organically to provide quality for everyone—and that includes the world's poorest.

To the many people who ask me how we are going to achieve all this, the simplest, most effective reply is summed up in the reality you stand for.

It's the *simplest* because what you do is virtuous in and of itself. What you do is what your fathers did before you. Keep on doing it. In one sense you're lucky: you don't have to invent anything new.

It's the *most effective* because what you do is also extremely tangible. Land, food, living creatures—this is real economy we're talking about. Things we can touch and taste, from which we can take pleasure.

It's all a matter of love and care for the local economy. You haven't got to do anything more than what you're doing already, aside from improving, boosting, and supplementing it by networking with one another. But, believe me, in many senses your labor is already the future. Within the context of your local community, your activities trigger positive effects. Above all, you promote traditional diets: healthy, tasty, varied meals that follow the seasons and provide plenty of nourishment.

If food was no longer obliged to make intercontinental journeys, but stayed part of a system in which it can be consumed over short distances, we would a save a lot of energy and carbon dioxide emissions. And just think of what we would save in ecological terms without long-distance transportation, refrigeration, packaging—which ends up on the garbage dump anyway—and storage, which steals time, space, and vast portions of nature and beauty. In a local economy, energy and resources are optimized and waste is avoided. Here in Italy alone, that waste means 4,000 tons of edible food thrown away every day. Seen through the eyes of people who suffer from hunger— but, come to that, also through those of rich countries— 4,000 tons is an absolute disgrace. Farmers everywhere have always reused and recycled waste; they throw nothing away, recovering refuse from any given process to produce more energy and make tools and artifacts. Waste can also be recycled—sometimes for gastronomic purposes— or transformed naturally into fertility for the soil. And as to water, farmers don't waste it; they store and respect it.

We can achieve all this by working at the local level. And by working at the local level, we can also achieve true participatory democracy. Today the whole world is hungry for participatory democracy: at the local level anyone can take part. Anyone can play a leading role and become an active partner.

At the local level, we defend biodiversity, a vital resource; for native animal breeds and vegetable varieties are the irreplaceable driving force behind small-scale agricultural economies.

The concept of "local" also translates into social relations. Today it's up to young people to help build a new rural sociality and conviviality. All of you young people among us, I urge you to read, study, and learn about the past; know who you are and where you come from; but, at the same time, try to promote social relations in the countryside and enjoy being there. Use the new tools available to you. Use cameras to film the know-how of old farmers and peasants. If you don't, it'll be lost for good.

We all know that this type of economy is vital if we want to have a better future. But that doesn't mean refusing to collaborate or sticking to the past. Your feet are firmly on the ground and your gaze is fixed on others, ready to catch a glimpse of anything that might improve the earth. Hence the importance of the network, your network, our network—a network made up of local food communities. Can you imagine a more open-minded solution than this? A lot of folks take a condescending view of the local dimension. They say your products are too minor, too marginal—but the opposite is true. If you join together as one, you small-scale producers arguably form the biggest food multinational of all. The difference is that you don't produce standardization or pollution or poverty. You produce wealth, diversity, exchange, preservation of memory, and progress. Here we have the value of the local economy.

Your economy is the most modern thing in the world today. To all those who, maybe in good faith, see Terra Madre as some sort of laid-back get-together of poor, marginalized human beings—as if you were a bunch of losers—I say: you just haven't got it. People who think like this just haven't cottoned to the fact that it's here that the future is being played out, that it's you who represent the swarming mass of the world's farming classes and villages—half the world's population!

Farmers and peasants will be the leading players of the third industrial revolution, which sets out from your villages, your businesses, and your land. The first industrial revolution started with the steam engine and the second with electricity; they both relied on energy derived from fossil fuels. The third industrial revolution will be the revolution of clean and sustainable energy. It will start in the countryside, agriculture being the only human activity based on photosynthesis. For centuries, the farmer's work depended on sunshine, and I exhort you now to develop and use clean, renewable energy. Produce solar and wind energy and energy from anything else you can come up with to generate wealth for your businesses and for your families. This is the "New Deal" for tomorrow. Contact anyone you know who is working in this direction. To the many managers in the audience who want to work in the field of sustainable energy, I say go and pick the brains of farmers and peasants. They know how to reuse matter, they have a close relationship with the land, they waste nothing. To all the rest of you I say you already possess the know-how it takes for the new revolution, the revolution that, to a large degree, will accompany the revival of the real economy.

You should be proud of yourselves. You are representatives of the world's diversity; you are the greatest wealth of all; our greatest resource, and the guarantee for the future of humanity. Just as biodiversity exists in nature, thereby ensuring survival, evolution, and adaptation, so your identity, your traditions, and your customs are something that the world can't do without. No, the world would be nowhere without your cultural diversity. It is precisely when our diversities meet and mix and connect that each one of us strengthens his or her own identity. If we were all equal, there would be no such thing as identity: identity exists thanks to diversity.

Enjoy the coming four days to the full. Don't be afraid to express yourselves. Don't have reservations about meeting and getting to know others; even though you speak different languages, even though you dress differently, even though you have different-colored skins.

Let's sit down and talk. Maybe you think you won't understand everything at first. But all it takes at times is a smile or a gesture or a handshake. Believe me, this is a unique opportunity for you to come to terms with the world's diversity; to find out and learn, but also to assert yourselves and the pride of your own identity. And when you go to the houses of the Piedmontese people who are hosting you, take the opportunity to get to know them. Exchange ideas, gifts, smiles, conviviality, and culture. Our past experience of the Terra Madre event has taught us that there's no such thing as a language barrier, that language isn't a problem. Don't worry, you'll have no problems in communicating; you will always understand each other, come what may.

Last but not least, take the spirit of this event back to your communities with you. Once you get home to your villages, open them up to welcome others. The young people present here today are about to launch a fantastic idea. They'd like to spend a few months of their lives working for nothing in your agricultural communities. Welcome them, open your doors to these young farmers and students. They have decided to devote two or three months of their lives to working for you. This will enrich you and enrich them. It will be the greatest exchange of youth and culture the world has ever seen.

My final words go to the young people among us. You are the future of the earth. Save the memory of farmers, save the memory of your elders. There can be no future without memory. Make your memory of the elders and farmers of your villages a cornerstone of the new frontier; allow the traditional wisdom of your elders to converse with modern science, and you will be the makers of your own future.

I'd like to say to Sam, the young man who preceded me, that he comes from a great country.[1] Maybe in a few days' time that great country, the United States of America, will give us new hope. Maybe a dream will come true that only a few thought possible.[2] But you know, Sam, even this hope isn't as strong as the final words of your speech to this meeting: "We will be the generation that reunites mankind with the earth."

If you raise the flag, we will be behind you in this revolution. Mankind needs to be reunited with Terra Madre, Mother Earth, and you young people can make this happen. So, Sam, when you go back to school, take with you the words of a great man and a great American, the Dakota Indian Red Cloud: "The earth appeals to humanity for redemption. This redemption lies in our common sense, in our rectitude. The earth waits for those who are capable of distinguishing its rhythms. Look at me: I am poor and naked, but I am the chief of the nation. We do not want riches, but we

---

1. Sam Levin, a young student and organizer of a school garden in Massachusetts, who spoke before me at the Terra Madre opening ceremony.

2. The reference is to Barack Obama, who won the U.S. presidential elections on November 4, 2008, a fortnight after my speech.

want to train our children right. Riches will do us no good. We could not take them with us to the other world. We do not want riches. We want peace and love."

With this, the message of a great man, a great American, a great American Indian, I invite you all to live the next four days of Terra Madre intensely, with joy and with passion. I hope you will keep these days in your eyes and in your memory. A happy Terra Madre to you all!

terra madre

ONE

all about terra madre

Terra Madre first appeared on the global political and economic scene in 2004. It began as a large meeting of people from all over the world, but soon turned into a permanent network—or rather a number of networks—whose members work day by day, wherever they happen to be, to create a new economic, agricultural, food, and cultural model.

Terra Madre is a concrete way of putting into practice what has been defined as "glocalism": a set of actions carried out on a local scale to generate major repercussions on a global scale. It has evolved over the course of time and now has a policy of its own, shared values, and medium- and long-term objectives.

Terra Madre is thus much more than just a biennial get-together (the organization of which is, I might add, a mammoth task). It is also a world network of local food communities, the living, breathing embodiment of the ideas I am about to set out. Let me begin by describing Terra Madre's distinctive features, its history, and the activities it seeks to and can perform.

Once you have understood what this great network is, the ideas and philosophy behind it will acquire a different flavor and a greater depth. Above all, you will appreciate how they are closely bound to the real world and to a heterogeneous mass of men and women, all working on a common project.

There is a lot of talk these days about the need for a new system of world governance. It is now clear that we cannot rely solely on the governments of the strongest nations and the potentates of the economy. We also have to make overtures to the political representatives of the countries of greatest demographic importance, nongovernmental organizations, and all the major large-scale movements that work in the areas of the environment, social justice, and volunteerism.

This is why I believe that Terra Madre, which will seek to increase and consolidate its political status over the next few years, stands for the future. It is an open network that welcomes anyone who shares its ideals, even if they do things differently or work in diverse geographical and operating contexts. It embodies a new approach to the production, processing, distribution, and consumption of food, drawing liberally on the history of the world's populations, but also looking ahead. It's conscious of the mess we have gotten ourselves into, but it's not afraid of the future.

How the Idea of a Terra Madre Network Was Born

"We've got to organize a large meeting of small-scale farmers, the people who work the land in every corner of the globe. Let's invite thousands of them to Turin. All we have to do is find the resources to get them here. They'll see to the rest, they'll provide the content. And I want everybody to be represented: not only Europeans and North Americans, but also people who live in the most far-flung villages in Africa and South America, people who have never left their fields in their lives before. What's important is that they have three things in common: that their labor helps save biodiversity, that they work in harmony with the earth, and that they produce good food."

It was a morning in late 2003 and I was talking with my closest collaborators. You should have seen their faces! Their mixed reactions were a perfect reflection of their different personalities. But all of them were shocked by the sheer number of people I intended to involve in the event. In an instant, they began speculating about the type and amount of work that was lying in store for them. The machine went swiftly into motion: some people set out to raise funds; others began to figure out how a meeting of this kind might work. We immediately came up with a name for the event: Terra Madre, in honor of Pachamáma, the South American Indian name for the Earth Mother venerated by millions of farmers and peasants all over the world.

My idea was inspired by the Slow Food Award for the Defense of Biodiversity, which, from 2000 to 2003, we had presented to farmers, fisherfolk, and food artisans whose daily labor had helped save small fragments of biodiversity: an animal breed, say, or a native plant variety or maybe a traditional way of turning nature into food. A dozen or so people received the award each year; they were chosen by a jury from a short list of hundreds of nominations put forward by a network of almost seven hundred journalists in about eighty different countries. These journalists—a diffuse network of professionals having the wherewithal to gather and evaluate stories of rural people and traditions and sensitive enough to evaluate their importance—formed the real backbone of the event. It was they who introduced us to a multitude of wonderful human beings whom we contacted and met during our travels overseas on Slow Food business.

Under the award regulations, this network of journalists had to meet at the award ceremony every four years, and organizing such a "happening" was no mean feat. The first Slow Food Award ceremony was held in Bologna in 2000, the second in Oporto in 2001, the third in Turin in 2002, and the fourth in Naples in 2003. The journalists all came together in a large plenary gathering at the first ceremony in 2000, but from 2001 to 2003 only the winners were present. In 2004 the time had come to organize the journalists' meeting again.

It occurred to me that, with the same outlay of funds and resources, instead of the journalists, we could invite thousands of farmers—who, after all, were the true heroes of their fantastic stories. We would be able to introduce them to one another, give them the chance to travel, let them see a different world, and make them feel pride in the hard but skilled jobs they do every day. Hence the idea I threw at my collaborators. I was confident that all we had to do was raise the funds to pay for travel expenses and take care of the logistics. I knew that the civic and religious groups and, above all, the farmers of Piedmont, my region of Italy, would be only too willing to help me arrange sleeping accommodations for our guests. There's no point in obliging farmers to stay in grand hotels, as if they were normal conference goers. These farmers possess a great sense of dignity, and they know how to adapt but, in many cases, they've never traveled before. Who knows, they might feel ill at ease in an anonymous hotel room. I felt sure they would be happier finding out about farming and talking to their counterparts around the Piedmont region than being seduced by the attractions of international hotel chains.

By utilizing the Slow Food Award journalists' network, the 80,000 members of our association worldwide, and the many food producers involved in our projects to defend biodiversity—such as the Ark of Taste and the presidia[1]—and after a painstaking selection process, we began to send out invitations. We chose October 20, 2004, as the date for the event so that it would coincide with the biennial Salone del Gusto (Salon of Taste).

Since 1996 we have staged the Salone, a major international event, at the Lingotto Fiere Exhibition Center in Turin. A showcase for the food of Italian and international producers who comply with the Slow Food philosophy, it has always been a huge success with the public. What better moment to invite along peasants and farmers—the hands and minds that are the creators behind the wonders to be found at the Salone del Gusto? And why not seize the opportunity to let them visit the Salone, a point of reference for international gastronomy? It would be an encounter between visitors and quality food producers and thousands of farmers from all over the world: the people who, through their toil—and despite difficulty, poverty, and some-

---

1. Slow Food presidia are small-scale projects designed to safeguard biodiversity and traditional food products at risk of extinction. There are currently 177 presidia in Italy and 121 in another 46 countries. The projects involve more than 10,000 small producers: farmers, fishermen, butchers, shepherds, dairypersons, bakers, confectioners, and so on. They are concrete, virtuous examples of a new model of agriculture based on quality, the recovery of traditional knowledge, respect for the seasons, and animal well-being. They support and promote quality food products that are: **good** to eat and typical; **clean,** made using sustainable techniques and with respect for the environment; and **fair,** made in conditions respectful of workers, their rights, and their cultures, and with a decent guaranteed wage. They reinforce local economies and help forge a strong alliance between producers and consumers. The presidia are promoted and coordinated by the nonprofit Slow Food Foundation for Biodiversity (www.slowfoodfoundation.com) and are part of the Terra Madre food community network.

times dramatic living conditions—manage to give us our daily food. It was a chance not to be missed, a chance to raise awareness of these people's problems and their fight for civilization.

It wasn't easy to organize the event. Imagine what it means to move 5,000 people, many of whom are not used to traveling, from sometimes remote villages, and book flights, arrange accommodations, and organize domestic transportation for them. In the end the meeting thankfully went off without major hitches and turned out to be a resounding success. With the help of national and local institutions, we raised the money needed to pay for the journeys of all those—the vast majority—who couldn't afford them. I was particularly moved by the way in which Piedmontese farming families were prepared to offer hospitality to delegates. On October 20, 2004, at the Palazzo del Lavoro fair hall in Turin, about 5,000 people from 130 countries met in plenary assembly, while just a stone's throw away, at the Lingotto Center, tens of thousands of people were simultaneously visiting the Salone del Gusto.

There were peasants and farmers present at the gathering, but also fishermen, artisans, shepherds, and nomads: no trade-union representatives, no association leaders—just people who work the land and sail the seas to produce food for all of us.

These people had come to Turin to represent their communities, which we decided to define as "food communities." Food communities could be groups of producers from the same place making the same or complementary products; groups born of alliances among growers and processors; or entire local food-supply chains. In short, the 5,000 present in Turin were representing a number of people at least a hundred times greater. It would only be stretching things slightly to say that they formed the biggest food multinational in the world.

Over the four days of Terra Madre 2004, the food communities present were encouraged by the likes of Charles, the prince of Wales, and Vandana Shiva, to celebrate together by day and in smaller groups by night in the Piedmontese villages where they were staying (and where they had a whale of a time sampling the food and wine and finding out about local agriculture). At the event, theme workshops addressed common problems and motivated participants to exchange their experiences. That's the only part of the event that Slow Food actually organized. What we attached most value to were the participants' journeys—the opportunity we were giving them to set out and discover a different part of the world and a different kind of agriculture—and their meeting together, which was a chance for them to represent their culture and feel pride in their own identity—identity being defined by differences. We knew that such a coming together of diverse people and cultures was going to have unpredictable results, so we confined ourselves to seeing that the community delegates were well looked after.

We weren't wrong in our assumptions, because the most wonderful things did happen. Though we had arranged for simultaneous interpreting in seven languages, farmers from different countries readily improvised conversations, gesticulating to explain their techniques, their crops, the characteristics of their land, and so on. And they understood each other! What better demonstration was there of the fact that no borders or barriers exist for people who love, know, and work the land? In many cases, the relationship between hosts and guests turned into genuine friendships. Many Piedmontese farmers have since gone to visit the communities of the people they hosted. Many others have kept in touch either out of friendship or because, albeit in different corners of the planet, they are engaged in similar battles, or because they need to seek information or follow experiments or exchange agricultural techniques. The network of friends that came into being went beyond our wildest dreams. Even before the various communities left Turin, we realized that we had a new global network on our hands, capable of acting locally but also of rallying as one: a collective made up of people who, until a short time before, had felt they had been left to fight a stoical but losing battle all on their own. A hope was born: in them, in us, and in all those who had seen us at work.

People who had said we were crazy for wanting to organize an event of this scale and scope were forced to eat humble pie. We too were pleasantly surprised to find ourselves involved in something that was even larger than we had imagined. True to form, our first thought was to up the ante: to involve more and more people and open up the network according to our holistic vision of the world of food: a vision in which agriculture, cooking, economics, history, and science subtly interconnect; in which different forms of knowledge, functions, and abilities cross over systemically; in which there is no place for reductionism and mechanism.

Thus in 2006 we enlarged the Terra Madre agenda to include culinary know-how and the dialectics between modern science and traditional knowledge. Besides the representatives of food communities, we invited 1,000 cooks—from Michelin-starred chefs to village innkeepers—and asked them to compare their experience with that of the people who produce food. By undertaking to use the produce of the communities closest to their establishments, to a certain degree these cooks became members of the communities themselves. In the United States, for example, an ever-increasing number of chefs and cooks are committed to using fresh, seasonal, sustainably produced ingredients. And, along with cooks, to Terra Madre 2006 there also came about 250 representatives of universities that had pledged to research and reappraise popular and traditional knowledge, to put old and new on the same level without prejudice. In the end, almost 9,000 people turned up in Turin, this time at the Oval, a structure built for the 2006 Winter Olympics and adjacent to the Lingotto Fiere Exhibition Center, again in conjunction with the Salone del Gusto. The opening ceremony was attended by Giorgio Napolitano, the president of Italy. The network was growing, and increased media attention was helping the public to appreciate its importance.

In 2008, Terra Madre's farmers, cooks, and academics were joined by rural musicians, inheritors of the oral and musical traditions that have always accompanied labor, the seasons, the rites of passage, and the convivial celebrations of the countryside. Music too is part of the world of food and peasant culture and embraces forms of diversity that deserve to be saved. In 2008 we also asked communities to bring their young members; for it is they who have to assume the huge responsibility of assimilating the knowledge handed down to them and continuing to tend the land of their ancestors. But it is also they who can provide hope for the future of the planet. Last but not least, we invited natural-fiber producers who, in the countryside, work side by side with food producers and address similar problems, such as soil erosion, competition from industrial products and manmade fibers, loss of biodiversity, and the inequitable laws of the market.

Today, after the third event, the Terra Madre network consists of about 1,600 communities from 153 countries. It now also includes the thousands of young people who have joined the Youth Food Movement, a groundswell that began in American universities and has since spread internationally. Their pledge is to improve cafeteria meals, to promote food systems education, and to encourage and support peers who plan to become farmers. The number of cooks and academics involved is growing, as is the geographical spread of the phenomenon. The network is further bolstered by Slow Food, whose members every day seek out food that is better for themselves and for the earth, and who are engaged in alternative forms of distribution such as the Earth Markets[2] or the creation of school gardens and other educational projects designed to defend biodiversity.

Not that the network will end here. We also intend to promote local meetings and education programs, as well as national Terra Madre events (several of which have already been staged over the last few years). In this way, we will strengthen the network and foster the circulation of ideas and constant media impact. Of central importance from this point of view was Terra Madre Day on December 10, 2009—the twentieth anniversary of the founding of Slow Food—when food communities celebrated Terra Madre on their home turf with small- and large-scale initiatives dedicated to the subjects of food sovereignty and the local economy. From now on, we also intend to involve community-supported agriculture (CSA) in North America and similar models of food distribution around the world—such as Gruppi

---

2. For more information about Earth Markets go to www.earthmarkets.net.

di acquisto solidale (GAS), or fair-trade shopping groups, in Italy, and the Association pour le Maintien d'une Agriculture Paysanne (AMAP) (Association for the Maintenance of Rural Agriculture) in France—on a more structured basis. In our holistic vision of the world of food, the life of rural and food communities is closely connected with, among other things, distribution, communication, memory, dress, quality of life, and building techniques. No attempt to enhance agro-gastronomic heritage can ignore the fact that all these elements are interlinked. Take language, for example. Some 5,900 indigenous languages spoken by 3 percent of the world's population—most of whom are members of our food communities—are at risk of disappearing. A language encapsulates a worldview, and its words describe traditions and methods of working, as well as tools, utensils, and natural phenomena. Like food and traditional ways of producing it—not to mention selling, celebrating, storytelling, and preserving collective memory—language is part and parcel of the cultural identity of food communities.

Through Lingua Madre (Mother Tongue), a project we are developing in conjunction with the Piedmont Regional Authority, we are striving to save ancestral languages, besides promoting the positive values of intergenerational collaboration, which values the wisdom of the older generation—an intellectual treasure trove that would otherwise be at risk of disappearing. It is the whole local system that needs to be protected, preserved, and held together, not just a part of it. And it's a system that gravitates around one huge sun: food.

Terra Madre: Poetry and Politics

As it has grown, Terra Madre has triggered mixed reactions from the outside world, while forcing us, its inventors, to address new problems every day. The decision to make the meeting coincide with the Salone del Gusto every two years has, to some extent, determined public and media perception of the event. Not that things have always gone as planned.

In 2004, the first Terra Madre event was closed to the general public. Only delegates were allowed to attend, and understandably so: organizers who transport 5,000 people to northern Italy to allow them to meet and work together (and I emphasize that they *do* actually work—Terra Madre is no feel-good photo op), can't be expected also to manage the general public.

But in 2006 we did decide to open the event as far as possible to interested observers, and moved it closer to the Oval, alongside the Lingotto Center, the venue of the Salone del Gusto. The choice was dictated by the fact that in 2004, contrary to our intentions, Terra Madre and the Salone del Gusto were perceived by many as separate events. The first was the meeting of the world's poor, farmers, and peasants in colorful, picturesque costumes whom Salone goers would bump into in the street and reporters would write about in the papers. By comparison, the Salone was seen as a sort of land of plenty, with good, high-quality food to be enjoyed by visitors who could afford it. This impression was of course the exact opposite of what we had in mind, and it failed to reflect what was really happening. We soon got the message: what we were up against was ingrained prejudice. The connection between food, gastronomy, and pleasure on the one hand, and agriculture, culture, and commitment on the other, could by no means be taken as a given. Far from it: large sections of the population couldn't get their heads around it at all.

True, when all the Terra Madre participants get together they are quite a sight—especially when you see them for the first time. Faced with such a melting pot of diversity, you may be taken aback. Right from the first year, many delegates decided to turn up proudly arrayed in their traditional costumes. The *coup d'oeil* when they met in plenary assembly or strolled around the Oval and Lingotto was incredible. What you were seeing was a real cross-section of the people who live in and work the world's fields, but you could have been excused for mistaking the get-together for a spectacular parade (the traditional dress of the poorest people are often more colorful and striking than that of American and European farmers, for example) or a nostalgic cocktail of noble sentiments, one part bucolic to one part mawkish. But no, that's not what Terra Madre is about.

Of course, seeing all these different people gathered together in a single space or wandering, happily and curiously, around the exhibition pavilions is a delightful aesthetic experience. It was a moving sight, and I'm sure that all those who witnessed it were stirred and motivated by it. This is the poetry of Terra Madre, inspired by the profound dignity and beauty of all these people, by the thrill of seeing them together or imagining them in their own environments. But the event is also about politics.

Terra Madre makes the dream of uniting the world's humble outsiders and enabling them to feel important come true. The real importance of these people has always been underestimated. Regarded as being backward, not in step with the times, even "underdeveloped," the small-scale farmers and other food producers were almost regarded as outcasts. But this negative judgment is based on a vital error, one that might prove fatal for the world's political and economic system.

The fact is that politics has excluded farmers and peasants from the democratic process. In the wealthy West this is because their number is almost irrelevant; in much of the world's South it is because they are considered as masses to be exploited at election time, but undeserving even of a decent quality of life. Nonetheless, these people have preserved wisdom and knowledge, continue to produce good food in harmony with the earth, and maintain values and practices that we in "developed" nations sacrificed on the altar of consumerism long ago.

The Terra Madre food communities will save what we erased: sustainability, sobriety, and a more human, real economy. The future of our food is in their hands. All they ask is to be listened to; to be allowed to continue according to their own traditions and aptitudes; to be respected.

Food is politics, respect for diversity is politics, the way in which we care for nature is politics—Terra Madre is politics. There is nothing belittling about the fact that this kind of politics is poetically nuanced, that here the beautiful and the noble fuse with the serious and the tangible. Ethics and aesthetics can no longer be kept separate. The poetry and politics of Terra Madre teach the world to stop brutalizing itself, to halt the process of global homogenization that is debasing people and depriving them of any power of self-determination.

Terra Madre is steeped in values that are revolutionary, that is, capable of changing our destiny, of transforming each one of us profoundly. There is no other way to prevent calamities from adversely affecting the planet we live on.

The Values of Terra Madre

The people of Terra Madre are humble but proud, and also a very mixed group. The knowledge they deal in is at once ancient and very modern. Outgoing and colorful as they are, they inspire both poetry and political commitment. First impressions risk being tainted by prejudice. It is virtually impossible to define them exactly, these people of Terra Madre. In no way can they be pigeonholed or catalogued. It would be wrong to judge by appearances, and this is why it is necessary to delve deeper. We need to figure out the values that drive these people. The truth is that, ultimately, they do not need to be defined, but they do have the right to stay as they are. Nobody discovered them; we only arranged for them to meet one another.

The Journey and the Meeting

The whole project was sparked by two simple concepts: the journey and the meeting. Most of the people in the Terra Madre communities had never been outside of their own villages, much less outside of their own countries; either because they didn't have the money or time to venture forth, or simply because nobody had ever asked them to go anywhere. When we conceived of Terra Madre as a meeting, our primary aim was to convey to these people that they weren't alone, that realities like their own were to be found all over the world, that they had allies. Any journey is educational, any journey opens the mind. So it is only fair that, in a globalized world, these people should have the opportunity to learn by experiencing and encountering other cultures.

The very fact that we were able to give them the chance to travel thus seemed to us to be a success in and of itself. An identity, after all, is fortified and defined more precisely by difference and exchange. Bringing all these people to Turin to speak about themselves and the Mother Earth they are so intimately bound to elevated the significance of the project as a whole. I still recall how moved I was in Addis Ababa in 2008 when I heard an old Ethiopian woman, a member of the Wenchi Volcano Honey Producers' Community, read a primitive but highly evocative poem at Terra Madre Ethiopia, one of the national events born in the wake of the Italian one. The poem she had written captured her spontaneous impressions of her trip to Turin, the first time she had ever left her village. In those monotone, repetitive verses, she expressed her thanks to all those who had given her the opportunity to see another nation, another land, another agriculture. Her poem was a mixture of wonder and fear at the discovery of such marked diversity, but it was also full of gratitude toward the Piedmontese farming family that had hosted her during her stay. For the first time in a life dedicated to producing food for her family, she had left her village. And the temptation to jot down her impressions on paper had been too strong to resist. The old woman's style was shaky, but it conveyed her sensations directly, with all the intensity and wonderment of a small child recounting her first breathtaking adventures in the world. It was hard not to be touched by her words as she spoke.

The effect that travel can have on these people is masterfully described in the movie *Seu Bené vai pra Italia* (Mister Bené Goes to Italy), directed by the Brazilian Manuel Carvalho. It tells the story of Benedito Batista da Silva, a sixty-year-old peasant and a veritable institution among manioc growers in the state of Pará, in Amazonia, and his journey from his village, Bragança, to Turin. In the documentary we see the meeting of different cultures and between small-scale farmers from all over the world and from throughout Europe. As the relationship between Seu Bené and the family that puts him up develops and turns into friendship and affection, so cultural, economic, and even physical barriers crumble. Without speaking a word of Italian, Seu Bené has a great time chatting and chancing upon new discoveries; he relishes the prospect of returning home and telling his fellow villagers all about what he has seen, how he got on, the emotions he felt. The film is tender yet powerful: it shows how, maybe just a little, Seu Bené's life has changed forever, as have the lives of those around him. From now on, he and his buddies will have more pride in what they do, will be aware that they are part of a world, however far away it is, that is working with them. They will realize that they have found new friends.

Travel (the journey) thus leads to encounter (the meeting); a nice way of asserting and expressing oneself, of understanding others, of learning from diversity or uniting by way of affinity. This has always been the most moving and motivating aspect of Terra Madre. The meeting has triggered the most significant initiatives and the most enduring liaisons among newborn producers' associations. In the meeting each identity is imbued with pride and gains awareness of its role as an important piece in the mosaic of Terra Madre. For delegates, meeting each other has above all meant no longer feeling isolated and alone. This is the opinion of Markus

Friedrich Schumacher of the Biodynamic Producers' Community in Tula, Russia. Interviewed in Ermanno Olmi's film *Terra Madre* (2009), he says, "I went to the meeting as a lone fighter, but when I came home I no longer felt like that. I felt as if I was part of a great movement. My aim hasn't changed that much. The only difference—and it's a big difference—is that I no longer feel I'm pursuing that aim on my own. Every day I get up and go to work, which I really love. This love I transmit to everything I do, and the main difference between our produce and conventional produce is the result of that love. You can see, feel, and taste this in the produce that ends up on tables."

Like any other meeting, Terra Madre arouses curiosity, with farmers, fisherfolk, and all the other participants always asking each other how they go about their work, solve given technical problems, get the best prices, irrigate the land, process food, breed livestock. The exchange of even the simplest, most basic information has been, and is, massive and totally unrestricted.

Self-Esteem

In my view, the most important thing delegates take home with them—and that convinces them to stay on with us as members of the network—is a great sense of self-esteem: a newfound positive opinion of their own worth and that of the work they do every day. This is no mean feat in a world like the present one, and for the thousands of Terra Madre delegates it appeared less attainable than for people who don't deal directly in food.

Imagine a small-scale farmer in a seemingly godforsaken village in Burkina Faso. Or a fisherman with a small boat on a remote island off the coast of Southeast Asia. Or a rice grower in

Madagascar. Or an Indio in Brazil's *cerrado*. Or a woman whose seed-saving work is helping to protect a part of India's incredible biodiversity. Or a nomad shepherd in Mongolia. Or a Sami following his herd of reindeer across Norway, Sweden, Finland, and Russia. Or a drover in the Abruzzo region of Italy. Or a dried-fruit producer in Afghanistan. Or a vegetable grower near Sarajevo, in Croatia. All people who work every day till they drop, who hardly have any time for themselves, who can never move away from their homes, who don't always earn a fair wage; but who apply themselves, converse with nature, make it yield, provide food for their community and for many others. For the rest of the world these people were—and are—marginal and unproductive. Nobody takes an interest in them, and outside their own communities they feel like outcasts. The temptation to quit is strong. The idea that more opportunities are to be had in towns and cities is strong also. Many believe that it might not be a bad idea after all to chase after the dazzling lights of consumerism—a way for a *nobody* to become a *somebody*, perhaps.

These people often don't realize that the real wasteland is to be found in the temples of consumerism, now bursting over with poverty-stricken pariahs who have lost all hope. They, in sharp contrast, are still *somebody*. They possess the know-how to feed themselves and others, and to interact with their own natural and cultural context. After visiting countries ravaged by poverty, speaking of the locals, many Westerners say something along the lines of, "Sure, they're poor, but they look happy enough." It's one of the clichés you hear most often from people coming home from journeys to the world's South. But insofar as it's a cliché, it's also bound to contain a grain of truth. By which I mean that the inhabitants of the South really have conserved something that we in the affluent West have lost; something that should be natural and innate; something these people don't even realize they possess. Just as we don't realize that we've lost it.

Terra Madre has taught them that what they are doing is vital, that their work as "intellectuals of the earth" (that's how we've defined them since the first Terra Madre event) is indispensable for the future of our planet. It has also made them see that they are not unique; that in faraway places there are other people just like them who used to feel every bit as abandoned as they did. Self-esteem should not be underestimated; it is the point of departure for a personal journey of resilience, perseverance, and self-improvement.

Affective Intelligence
The recovery of self-esteem and the experience of the meeting not only enrich and motivate the food communities but also help to interconnect them. The communities have realized that, despite their differences, they have much to share. It's all a matter of aptitude, of a personal and collective history that has developed in a specific area of Mother Earth. The communities now know that they share a common destiny, that they are, as Edgar Morin put it, "communities of destiny" and are now masters of their own futures.

A sentiment is coming into being that I refer to as *affective intelligence*. It escapes the rules of consumer societies, in which individualism and the most rigorous rationality rule. The Terra Madre communities have always practiced fraternity, a sentiment guided by their wisdom, by their abilities, by their way of living life.

Of the three values of the French Revolution—*liberté*, *égalité*, *fraternité*—it seems that only the first two have survived. Albeit often invoked irrelevantly, they are widely recognized. But brotherhood, *fraternité*, seems to have disappeared from the order of values of Western societies. This is not true, it seems to me, of the communities of Terra Madre, and of the people who combine to form them, people who refuse to accept the rules of consumerism, waste, techno-mania, and industrial homogenization—mainly because they don't know what these things are.

Affective intelligence is the cement that holds Terra Madre together, and no external force can make it come unglued. In a world full of rational intelligence, at last we have a network that is affective and fraternal. You can breathe this in at the meetings of Terra Madre, or at its so-called Earth Workshops, but we also get a sense of it in reports that come in from regional Terra Madre events that are held around the world. It is something that came into being almost spontaneously at the grassroots level.

To date, Terra Madre events have been staged in Ethiopia, Tanzania, Sweden, Ireland, Belarus, Brazil, the Netherlands, Argentina, and Kenya. The network exists and makes itself felt locally and nationally, involving politicians by asking them to intervene in support of small-scale food production and the economies of local communities. It is this fraternity—this cement—that allows Terra Madre to grow freely without being overstructured and communities to achieve their own sovereignty and self-determination without outside imposition; to have fun while working hard and seeking their future with "no power, a bit of knowledge, a bit of wisdom, and as much flavor as possible."[3] Who better than the food communities can restore flavor to the earth?

---

3. Roland Barthes, *Leçon inaugurale* (Paris: Collège de France, 1977).

The Power of Diversity

Another positive thing about the Terra Madre communities when they meet is that no one tells them what to do. Certainly no one forces them to do anything. The guiding principle is that, "You are about to become the leading players in the third industrial revolution, the leading players in a 'New Deal' that will change the world and save us from the crisis we are now suffering."

Food communities realize that they don't have to invent anything; all they have to do is go on doing what they've always done. Nobody else is better than they are at doing their jobs, shaped as they are by centuries of local adaptation. The communities know how to produce food well, they know how to work the land, and they know how to reap its fruits without harming it. They reuse and recycle, and they use waste to improve soil fertility and thus to increase both the quality and quantity of food produced. They have always operated systemically, and this is their productive revolution.

All outside players can learn from the community model. They can use it to adapt and change their own systems, applying what they learn even to totally different urban contexts to develop another style of living and working. It's all about their way of thinking and their methods of problem solving.

If you don't have a clear idea of the communities' intellectual and cultural scope, of their way of addressing so many aspects of life, you are liable to commit the typical error of colonialists: that is, you steal their ideas and resources, while at the same time trying to persuade *them* to change. This is not the attitude of the organizers of Terra Madre and those who support the community network.

It would be all too easy to put the Slow Food cap on Terra Madre's head and pretend that it is a Slow Food network. But this is not the case at all. Slow Food was only the catalyst, the leavening, if you will, that caused Terra Madre to ferment and rise. Through its members, projects, and interactions with the food world, it is Slow Food that is part of the Terra Madre network, and not vice versa. Slow Food provides support and resources, supplies knowledge, and ensures that the nodes work. But no one can claim it rules the network. That is why the Terra Madre network—or networks—is governed by what I call "austere anarchy," generated by faith in others, in diversity, and in nature.

This austere anarchy gives us hope; it teaches us to be unafraid in the face of the uncertainty that marks the future. We know that diversity brings with it creativity, evolution, and problem solving. We know that, if we respect her, Nature, the earth, is a kind and generous "mother." We know that others can help us, if they live happily and proudly, and are gratified by their labor.

Terra Madre may be a political subject, but no one gives orders. The communities know what they are doing and will continue to do it. The linear "business-as-usual" approach gives way to as much complexity as is possible. Instead of talking, we have to learn to listen. To see communities as something to be used would be pointless. Luckily, all these wonderful human beings couldn't be controlled, even if we wanted to control them.

It is easy to think of these people as a silent, hidden army, called to attention every two years to parade before the world. The knowledge, beauty, and culture they bring are a miracle beyond belief. The communities aren't there to be explained; they should be listened to and allowed to take their course. They represent the opposite of homogenization, of consumerism, of what Slow Food movement calls *fast life*—a lifestyle we don't like and have opposed ever since the movement was born. They stand as proof that not everybody has been tainted by the evils of mass production and mass consumerism—that there is still hope.

Their ability to interact fruitfully and respectfully with nature makes them repositories of forms of knowledge that not even the most sophisticated researchers, the most unprincipled financiers, or the most powerful industrialists can overlook. With their simple know-how, the people of the communities produce food, energy, and tools. And, in doing so, they also produce happiness, genuine human relations, and full and satisfactory existences for all.

The quintessential attribute of Terra Madre, the one that makes us have faith in those who are part of it and in the future, is diversity. Diversity stretches complexity to the bounds of incomprehensibility, preventing reduction to linear order and the application of economic theories that still have the upper hand but are clearly and increasingly incapable of offering us any real prospects. Diversity may appear untidy, unmanageable, and unaccountable, yet all we have to do is place our trust in it, allow it to grow, "listen" to parts of it, let it express itself and take root with the new generations, while respecting those of the past. Diversity is a celebration of intelligence and human creativity, of everything we ought to achieve to ensure that everyone has the right to eat enough of the food that they prefer, food that nourishes the body and the spirit. It is our hope that, by embracing and defending this diversity, we can achieve everything we need to prevent the earth from falling apart. To this end, complexity is not an obstacle but an opportunity.

That all those faces, that music, those hands are still out there on the land, in the communities every day, even as I write and as you read—this is the greatest element of hope for the future. These people are our guarantors; their guarantee, of continuing to be free, lies in the fact that their network is and will remain austerely anarchic. The network embodies diversity to the nth power, but this is no cause for concern, since these people are fully committed to working as food communities.

TWO

food communities

The food community is the leading player in a new global food policy. In this chapter I explain how the concept came into being as part of the Terra Madre phenomenon, and how it has developed since.

We coined the term "food community" toward the end of 2003, when we were looking for a common denominator for the people we were inviting to the first Terra Madre event in Turin. It was a way of describing a sort of "unit" from which we would pick the most charismatic representatives and the ones with the best communication skills (including knowledge of foreign languages)—in other words, the individuals best suited to coming to Turin for the meeting.

The short lists we were receiving, and the information that we already had gathered, painted an extremely variegated social and economic picture: villages, ethnic groups, small-scale producers' associations, families, consumer cooperatives, entire local production systems, sustainable food-supply chains, local and national urban farming movements. Geography came to our aid in the many cases where people lived in well-defined areas, and were thus easy to group together. But broader geographical areas were also represented, and sometimes the connection between the various players in a given community was only virtual or remote. But still, the term "food community" was effective, insofar as the people in question were members of groups that, however heterogeneous, shared common values and, above all, shaped their lives around food production.

With the staging of the three Terra Madre events and the consolidation of the permanent network, the meaning of "food community" has progressively broadened. If at first the term was mainly instrumental, now it has taken on theoretical connotations. Today the "food community" is the ideal, tangible pivot round which the philosophy and actions of Terra Madre revolve.

The original Terra Madre idea had envisaged the inclusion of food-producing communities alone. Adopting a holistic and systemic approach, the second event in 2006 expanded the "food community" concept to embrace cooks, academics, musicians, and producers of natural fibers as well—not to mention consumers, who we recognize as the allies of small-scale producers.

At the same time, we realized that the word "consumer" itself had become bankrupt of meaning. So instead we began to call the people who eat the food of the communities "coproducers." The fact is that their refusal to accept the consumerism of the global agri-food system means that they can no longer be classified as mere "consumers." These people are aware that "eating is an agricultural act,"[1] and they reward food producers who display a real feeling for gastronomy.[2] Conscious of the fact that their food choices have an impact on the general sustainability of produc-

---

1. Wendell Berry, "The Pleasures of Eating," in *What Are People For?* (San Francisco: North Point Press, 1990).

2. Carlo Petrini, *Slow Food Nation: Why Our Food Should Be Good, Clean, and Fair* (New York: Rizzoli, 2007).

tion processes, they refuse to lie down and be rolled over by the global agribusiness machine. They buy the food they need to feed themselves with great thought and intelligence. They know that every foodstuff is the end result of a network and that they too can be active players in that network. They see the final act of eating as the point of arrival of a cyclical process of production that itself is a cog in a much larger wheel and a much larger biological system—namely, nature. The term "consumption" is reductive since it implies an action that is at once sustained by the consumer, passive toward the food system, and harmful to the planet, which is what, in the end, is being "consumed."

On the contrary, coproducing—being a coproducer—means becoming part of a food community together with growers, breeders, processors, and distributors. It means carrying out virtuous actions together and believing in the rebirth of a food system that, once restored to harmony and balance, will enable the earth to prosper and regenerate.

Food communities are thus highly structured entities, made up of subsystems that, though limited, may be complex. They are firmly rooted in and look after the area in which their members live. There are no limits to the types of people who can be members, so long as they are concerned—at whatever level they represent—with the idea of good, clean, and fair food. Such people have an identity that preserves the memory of the past and at the same time expresses a clear idea of what they want for the future.

The time is ripe for a collective assumption of responsibility and a sharing of values that will make us all feel somehow like food producers. When they espouse the cause of good, sustainable, and fair food production, consumers change their role. The difference between town and country dwellers disappears. Even if we aren't farmers, we all need to start feeling like farmers again.

Communities Worldwide

It's one thing to speak about food communities on a theoretical level; it's another to know what goes on inside of them. To find out, I decided to extract a few cases at random from the online database (www.terramadre.org) of the approximately two thousand food communities in 153 countries that have taken part in the three Terra Madre events organized to date. Riveting and heroic, the stories that emerge reveal just how various and virtuous the communities really are.

Let's go to the town of Cotonou, in the commercial zone of Benin's Département du Littoral. Here, since 1992, a community has been developed around a family-run retail shop, thanks to a network of suppliers who stock ecological and traditional foodstuffs from more than ten thousand farmers, breeders, processors, cooks, and educational farms. Its products—which include honey, peanut and palm oils, karité (shea) butter, cashew and pineapple syrups, néré sauce, soy cookies, dried baobab and okra, and yam, manioc, white corn, sorghum, and soy flours—are bought by at least five hundred people every week. Many of these foods are commercially labeled to indicate place of origin. This may seem a minor detail, but it says a good deal about the community's conception of, and respect for, the foodstuffs in question. One of these is *zomi*, a flavored palm oil that is a specialty of the Adja women of the Mono region of Benin, and which is used to dress *niebé*, a local white bean, or boiled yam and manioc.

Still in Africa, in northern Cameroon, just south of Lake Chad, in the corridor formed by the northern territories that divide Nigeria and Chad, another community lives mainly off livestock breeding and small-scale agriculture. It is famous for its pigs, which are raised free-range according to traditional methods. After the pigs are slaughtered the members of the community cut, wash, and process the meat either for immediate consumption or preservation. The traditional dishes they make with it are *hliuma tchoufina*, pork grilled in a traditional oven, and *dohonokgna*, chopped pork stewed in a pan with *oseille* (sorrel leaves) and peanuts, and served with millet or rice.

In Congo Brazzaville, the Association Congolaise des Jeunes Cuisiniers (ACJC) was formed in 2002 by a young cooks' collective that brings together cooks, legal scholars, journalists, and students to promote Congolese cooking, produce, and foodways at home and abroad. Through its activities—free and professional cooking classes, television programs, local cooking contests, lectures, and debates—the association promotes native vegetable varieties and traditional cooking methods, while at once raising the cultural profile of the cooking profession. One of the typical dishes it is attempting to save is *liboké*, or *maboké*, in which the leaves of specific forest or savannah plants serve as natural containers in which various other ingredients are cooked.

Who ever said that Africa is poor and backward? Who says that it has no agriculture worthy of the name? In Ngombi, in Gabon, for example, a small community of fifteen young men and women grow, promote, and market local vegetable varieties, especially pumpkins—*ilengue* or *dilegue* in the local dialect—and cucumbers, or *teri*. The Ngombi project may not radically alter the gross domestic product of Gabon, but it does make eating easier.

African local agriculture is also represented within Terra Madre. In Ghana, for example, a community of forty young farmers in the district of Accra, all within, on average, ten miles from the capital, grow three varieties of *bankye* (cassava), each with its own distinctive characteristics: *abasafitaa*, ideal for the preparation of *fufu*, a thick porridge; *afsifi*, which is starchier; and *glemo duade*, the most watery. Many local recipes require the use of cassava: the abovementioned *fufu*; *yakayaka* (a sort of bread); *gari* (a fermented, granulated flour); and *agbelikakro*, or plantain fritters. Cassava may also be processed into flour to bake bread. Together with this staple, the growers also raise vegetables such as eggplants, peppers, tomatoes, and onions. They all use natural methods thereby preserving traditional production and distribution systems. The absence of chemical fertilizers ensures that the vegetables achieve the right degree of ripeness slowly and naturally before being sold directly or used to feed the growers' families.

Switching continents, in Argentina we come to an activity that is both productive and educational. The Escuela Rural No. 92 "Wolf Scholnik" is situated in the Rio Negro province, in the Andes almost on the border with Chile, in a village of eighty-five families of farmers and meat- and dairy-cow breeders. Here a number of vegetable growers began to collaborate with the village school to produce organic vegetables for its cafeteria. The running of the school garden has tangible as well as educational ends; pupils learn to produce food with respect for the environment and repeat the same practices in their gardens at home.

Some people resort to agriculture as a way of surviving in a dignified manner within conflict zones. Take the community of fruit and vegetable growers of al-Khadra, in the northwest Iraqi region of Ninawa, one of the areas hardest hit by the recent war. It is made up of thirty families who have joined forces to defend the region's wealth of traditional crops and biodiversity. Partly on account of the war, its members have discovered the value of seed saving through exchange, as well as the advantages of returning to traditional crops. The al-Khadra community uses only traditional seeds and grows its crops organically. Members of the community have organized to sell their products at the local market, and this has allowed them to diversify their activities and sources of income.

"And what of the "rich" countries? The United States is a breeding ground for interesting communities, such as the community-based garden project at Concord Commons in Rockford, Illinois. The garden is a project of the Angelic Organics Learning Center. In 1998, the biodynamic farmer John Peterson and a group of residents from the Chicago area founded the Angelic Organics Learning Center as the non-profit educational arm of Peterson's Angelic Organics Farm. His farm is currently one of the largest examples of community-supported agriculture in the country, with over fourteen hundred families from northern Illinois and southern Wisconsin participating in the weekly shares. The Learning Center has several initiatives that work to expand the farm's already considerable reach into underserved urban communities and further build local communities' connections (and access) to their food. One of these community-based projects is the Hands Working Together Community Garden at Concord

Commons, planted in 2005 as part of a joint educational program with Concord College. In the garden, parents and children from the neighborhood work as volunteers, learning all about how to grow food, while also learning about nutrition, team work and building a food community. The program furnishes technical assistance for organic gardening and youth education. This involves kids meeting with the coordinator of the volunteers and the staff once a week to plan gardening activities. Chores such as sowing, manuring, watering, and picking and gathering are performed directly by the kids. Fruit and vegetables from the garden are distributed among people in the neighborhood, served at dinners with workers and their families, or sold at the local farmers' market. In addition, kids have the opportunity to visit Angelic Organics Farm."

A model farmers' market is the one at Healdsburg in Sonoma County, California, where small local farmers are given the chance to come into direct contact with consumers. Though this area is dominated by vineyards and wineries, the farmers display a rich variety of produce at their stalls, demonstrating to the community that the local soil is capable of yielding much more than just grapes.

In growing their fruit and vegetables, these Sonoma farmers seek to achieve the best yields possible, while improving the health of the soil and enhancing the environment. Recourse to outside resources is reduced to a bare minimum. Among the fruit they pick are four varieties of figs from old trees cultivated by dry farming in a number of different localities. The only maintenance the trees require is pruning, and their figs are much appreciated by local restaurateurs and shoppers at the market.

Lack of space prevents me from taking into consideration entire continents such as Australia, or vast areas such as southern

Asia, Japan, and Central America, but I do wish to complete this brief overview with a final example from Europe, a combination of production and education. Parainen is a town in southern Finland that used to be the "capital" of a bilingual area where, though the most common language is Swedish, the population is still tied to its Finnish roots, especially to the tradition of baking. Bread, *leipä* in Swedish, is one of the pillars of Finnish food, and the Axo family bakery still keeps the tradition alive. Its specialties are black bread and *pullaa*, a light bread with a sweetish flavor, made with corn spiced with cardamom and cinnamon and decorated with almonds and sugar. The bakery has supported the "Grains for School" project whose aim is to serve healthy, flavorful baked products in school cafeterias to increase pupils' consumption of grains. This educational project involves not only the school kids, but also their parents, school staff, teachers, and local authorities.

Town or Country?

We have seen how a food community is much more than just a group of farmers or small-scale food producers from a remote rural area. Of course, such people are fully entitled to be members of the network, and are an important reminder that other lifestyles are possible, that traditional techniques and popular wisdom cannot be abandoned lightly. They perform an important function in preserving diversity—biological as well as cultural—and keeping alive modes of production that are vital for anyone seeking sustainable alternatives within the food system. Such people are without doubt the most "characteristic" and symbolic members of the network, but they are only one part—by no means the majority—of the multitude who form it. It would be reductive to limit the food-community concept solely to them.

To be, or to aspire to be, a food community is to attribute central importance to food, something that is possible everywhere. The concept applies not only to farmers, but also to consumer groups and cooperatives, such as AMAP in France, CSAs in the United States, and GAS in Italy.

These consumer groups forge alliances with farmers, and through various forms of payment (a deposit on all produce, a sort of "subscription" for the weekly delivery of a box of fruit and vegetables, or direct pickup and purchase at the farm), they circumvent or "leapfrog" the conventional market-distribution system to ensure not only a decent wage for producers, but also supplies of fruit and vegetables, meat, eggs, cheese, and even flowers to consumers, who thus become coproducers. All products are fresh and tasty and are entirely reliable because we know exactly who made or grew or raised them. Farmers and citizens thus get to know each other, make friends, and teach their children to respect ecology, economics, and culture—not to mention healthy human relations. What is all this if not a food community?

Communities also include horticulturalists who grow fruits and vegetables for educational and productive purposes: in the countryside, in schools, in town and city centers, and in suburbs. When Michelle Obama planted her vegetable garden at the White House just months after her husband had been sworn in as the U.S. president, she performed a very important political act. Planting a garden, even a small one, on the roof of a building or on a terrace has a high symbolic value. It means reappropriating food, food production, and the rhythms of nature. Insofar as it teaches us how to grow food and understand all the problems this entails, it means taking care of the land. In some urban farming and educational communities—in Australia, in the poor neighborhoods of San Francisco, virtually all over Italy, and in some African capitals—retirees go out of their way to teach schoolchildren the fundamentals of agriculture.

Even Yale University is getting in on the act. Its Sustainable Food Project was founded in 2001 in the belief that the world's health, cultural, environmental, educational, and economic problems cannot be addressed without due consideration of the way we eat and produce food. The project includes a farm on which sustainable agricultural methods are practiced, and where Yale students and the local community are educated about sustainable agriculture and the pleasure of sharing good work and good food. The project's ultimate aim is to train a new generation of leaders who perceive how people, the land, and food are connected, and for whom ecology, culture, and taste are more important than power or money.

Similar initiatives to the one at Yale have also been underway for some time on other American college campuses, often prompted by the efforts of students themselves. One example is the University of New Hampshire, where several years ago students held an organizational meeting to form an organic gardening club, and no fewer than three hundred interested people signed up. With the university's blessing, these students commandeered a little-used athletic field and planted vegetables, which were sold locally, including to the university dining services.

I cite these examples to demonstrate that it is wrong to think that a fully functioning food community can exist only in rural areas. In other words, what I'm suggesting is not a collective flight from towns and cities, but rather a different approach to food. Since, if we want to save the world, we'll have to do it together, what I'm fighting for is a new relationship between town and country. We can't leave all the responsibilities to farmers alone. The history of food has always been characterized by an alliance between urban and rural in which relations were close, systemic, and localized. Now that some of the world's metropolises are bursting at the seams, it's absurd that this alliance is falling apart too. We need to find new ways of cultivating and distributing food, of siting agriculturally productive activities, of allowing both town and country dwellers to draw sustainable benefits from the alliance.

Ways of doing this exist already, and new ones are being invented all the time. Looking ahead to 2015, the year in which Milan will host the Universal Expo, Slow Food, in conjunction with the University of Gastronomic Studies in Pollenzo in northern Italy (founded in 2004 by Slow Food in conjunction with the regional authorities of Piedmont and Emilia-Romagna) and Milan Polytechnic, has developed a complex project to radically redefine the relationship between city and suburban agriculture. Given the theme of Expo 2015—"Feeding the Planet: Energies for Life"—the idea is to make Milan a model for other urban centers to emulate, meaning that the city will at once host the event and transform itself as a function of its theme. We intend to practice what we preach—no more, no less.

Milan boasts an incredible resource: the Parco Agricolo Sud, an "agricultural park" that forms a 47,000-hectare half circle on the south side of town. Most of the area is intensively and conventionally cultivated, though 34 percent is also dedicated to perennial polyculture and eco-compatible systems. Only a tiny part of its agriculture, however, is at the service of the city and its inhabitants. Today, citizens of Milan in search of quality produce can only source it from outside the city, and it often reaches them courtesy of door-to-door deliveries from other regions. Since suburban agricultural land is being overbuilt, it is imperative to transform the farms of the Parco Agricolo Sud, make them profitable, and place them at the service of producers and citizens according to concepts of sustainability and innovation. As I write, the project is still at a preliminary stage, but it does seem to me to be an object lesson in the redefinition of the relationship between town and country.

Its aims are to identify and enhance existing good agricultural practices; implement innovative producer-citizen services; extend ties between town and country; activate permanent education projects; make an in-depth analysis of the whole project by way of the application of an ethnographic model; reconstruct supply chains through collective catering, inns, small-scale commerce, hospitals, and bars; promote the production of renewable energy; facilitate fundraising (land purchases, the rehabilitation and recovery of properties, and so on) to encourage young people to take up agriculture; and convey an image of the park and render it accessible as a landmark in celebration of the Expo (much like the Eiffel Tower for Paris). It is an ambitious project and all the points listed above demand specific courses of action, among which is the creation of a weekly Earth Market.

All food communities are excellent workshops for the redefinition of the town-country relationship. It is wrong to generalize—

every place has its own distinctive characteristics and needs—but the diversity that communities are capable of expressing certainly provides a rich, fertile humus in which new practices and habits can be grown. For too long we have believed that farming could exist without farmers, and that cities and towns—even the smallest—could let industry and large-scale distribution see to the rest. We have now understood, at our expense, and with serious damage to our ecosystems, that the models of standardized production and large-scale monoculture fail to solve the problems of food.

We have to relinquish our bucolic, anachronistic vision of the farmer-cum-food producer. We have to restore him his dignity and drag him out of the niche of economic backwardness and marginality into which he has been relegated. And we can do this anywhere on the planet, thanks to the alliance between producers and coproducers within the context of community, where genuine food production, interpersonal relationships, and knowledge transmission still count.

Parishes

The ongoing economic and environmental crises cry out for deep-reaching change. We are living through an age of transition, in which political, cultural, and economic management systems demand new paradigms. The analogy with the fall of the Roman Empire is by no means far-fetched. A system that had dominated the Mediterranean, most of Europe, and the Middle East was slowly deprived of authority. The collapse was by no means sudden: anything but—it took three centuries. Through that period, *pievi*, or parishes, areas of local autonomy, sprang up in every corner of the Empire. In these parishes, a new culture, new rules, new ways of understanding civilization developed independently. Parishes progressively grew into territories, some of them huge in size. They developed most, of course, where central authority was weakest, and it is interesting to note how the word derives from the Latin *plebs*, "common people." The development of community life required a new bottom-up organization; insofar as they all depended on the Church, it is possible to say that parishes formed a de facto network.

Analogies with food communities abound. The humility, immediacy, and concreteness of the communities, for example, are a response to the signs of decadence that are showing in the global agribusiness empire. The top five and a few other multinationals have almost managed to monopolize the ownership of seeds, land, breeding farms, and animal and vegetable varieties (the number of which is decreasing all the time) and are now proposing GMOs as the "final solution" to set the seal on their supremacy.

But after a century of undisputed expansion and interference at the local level—with consequent damage to biodiversity, local cultures, and the economic and ecological equilibria of the planet—cracks are now appearing in the "empire." The ongoing crisis has brought things to a standstill that raises important questions about the multinationals' actions. Thanks to their control over food, these global corporations are among the most powerful players on the world stage. Terra Madre communities combine to form a healthy alternative to the system, outside the corporate players' markets, far removed from their idea of economy, food, and agriculture. While the multinationals struggle in vain to find new products to allow them to maintain their dominance over global markets, food communities, like parishes, ignore their diktats and go their own bottom-up way, with the "plebs" and thanks to the "plebs"—a term hereby restored to its original, non-derogatory meaning.

In order to hand down ancient wisdom and produce food sustainably, communities make full use of memory and innovation (the Internet, for instance) without degenerating. They handle technology without becoming slaves to it; they make use of the good side of globalization to make themselves heard, to pluck up courage, to realize that they aren't alone in the world. They are a network that shares a common store of values.

The parishes of the new millennium allow themselves to be guided by Nature without claiming to dominate it. By way of the simple act of producing, processing, and eating their own food, they are actually performing an operation with a much profounder significance: they are restoring significance to the existence of human beings insofar as they form a part of Terra Madre. Their awareness makes food communities the ideal home for a new humanism, in which ethics and aesthetics blend as one, and in which the real and participatory commitment of individuals is predicated in a collective context both locally and globally.

Back to Basics: "Man Eats Food"

One day, after a screening of *Terra Madre*, Ermanno Olmi's documentary on food communities, their network, and the 2006 edition of the event in Turin, I had the chance to talk with the maestro.

We agreed that we don't know anything about food anymore, that maybe we need to go back to school to pick up the basics again.

"It's like revisiting parsing to understand the meaning of a sentence," said Olmi. "Subject, predicate, object. We've got to start from scratch."

Let's follow Olmi's advice and parse a simple sentence that lays the basis for our survival: "We eat food." It may appear banal to analyze such a simple sentence, but it is precisely because the sentence is simple that nobody stops to consider its meaning anymore. In reality, it succumbs to the paradox of the global food system. Slowly but surely, the active verb form has turned passive: "We are eaten by food."

Food is eating up the earth, its resources, and its chances of renewal. And since we are not a foreign body but one of the many elements of nature, we too are being eaten by food. As "consumers" of food, we allow ourselves to be eaten by it. As Zygmunt Bauman explains, "Anyone who is a member of a consumer society is in turn a consumer product."[3]

Consumerism is an ideology that pillages and wastes resources, but ultimately fails to satisfy needs. In the world of industrial food, this ideology has reached its apex. We ourselves are consumer products and, as such, can be robbed of our souls, treated like disposable goods, used without ever achieving a state of real well-being. As we lose every possibility of being active, we are gradually becoming the object of the sentence I wrote above. Food is eating us.

---

3. Of all the many analyses that have been made of consumer society, I find Zygmunt Bauman's one of the most exhaustive and profound. In his book *Consuming Life* (Cambridge: Polity Press, 2007), he cogently describes the harmful effects consumerism has had on people and how it has helped to increase our lingering sense of uncertainty. Political apathy and the rise in crime are just two of the effects that Bauman pinpoints.

Yet food communities, outside the system, are active. They are the real subject of our parsing. They eat food and don't allow themselves to be eaten. They can be producers and coproducers of food. They can be masters of a process that gives and ensures life, but they are respectful masters. They don't give orders so much as govern. The nice thing, almost revolutionary in its simplicity, is that any of us can be a member of a food community. And often we already are without realizing it. Or else a food community may be very close to us and can't wait to welcome us.

We have to become active subjects of the sentence "We eat food" again. We have to undertake to govern this planet of ours. We have to go back to basics and learn to be coproducers.

Here I have analyzed the sentence "We eat food" to try to work out who is or ought to be the subject and what characteristics he or she should have. In the next chapter, I continue the analysis with a special eye on the verb "to eat." Eating is an act that gives us pleasure, but increasingly it generates anxiety. We like to talk about it, even if we don't know enough about it. It refers to an act that, in some parts of the world, is taken for granted, whereas in others it signifies a question of life or death. It obsesses us more than ever before—and not only because food is running scarce. Eating has become one of the great paradoxes of our time.

THREE

food: dr. jekyll and mr. hyde?

If I want to eat well, I'm an elitist. If I respect tradition, I'm glued to the past. If I obey the rules of sound ecology, I'm a bore. If I acknowledge the importance of the rural world, I yearn for bucolic pleasures.

It's hard to talk about food and agriculture, sustainable food production and consumption without having to bear the brunt of this kind of cliché-ridden criticism. At Slow Food, we have noted that, now that things are starting to move, this kind of talk is rising in volume and frequency. For a whole variety of reasons, eating is increasingly in our thoughts and in our conversation. But instead of bringing pleasure and joy as it ought to, it generates uncertainty, unrest, anxiety, and fear. Eating, without which we can't survive, is turning into a problem.

In the world of today, the act of eating is pregnant with paradox. World hunger and malnutrition and the pandemics of obesity and diabetes are all sides of the same coin. We demand quality, yet we complain because it costs money. Then we go and spend the same amount on junk food or trashy consumer products. We watch TV programs that churn out recipes all day long, yet we've forgotten how to cook. We have all the food we could ever wish for at our disposal, yet we sweat buckets to slim down. At the same time, those of us who fight to protect fauna and flora on the verge of extinction, to promote the goodness that our countryside still has to offer, and to educate others about the pleasures of food are written off as elitist. It's as if, for cultural and economic reasons, it were no longer possible to combine pleasure and commitment.

How have we allowed this to come about? Food is our link with the outside world and nature: eating it makes us part of the complex system that the ancients described as "the breath of the earth." Metabolism is what distinguishes living beings from inanimate objects. We have a metabolism, what we eat has a metabolism, the earth has a metabolism, and all vital processes are closely interconnected. Arguably, at the root of the problem is the development model that has had the upper hand in all human activities, the eating of food being no exception. Industrialization and the primacy of a reductionist-mechanistic vision have heralded the triumph of consumerism. We have evolved into *Homo consumens*.

Human beings have convinced themselves they live outside the natural cycle. They think they can use nature to suit their whims. By virtue of their confidence in their own ability to produce anything, they think that even forms of production most closely bound to the natural world can follow the same laws. Once a badge of identity—a miracle of nature turned element of culture—food has now become a product like any other and, as such, complies with the laws of consumerism: meaning the laws of the market and of waste.

Our common store of practical knowledge—traditional and ancestral wisdom and the capacity to live in tune with nature—

has suddenly been erased and forgotten, as if the baby had been thrown out with the bathwater. Not that the cultural heritage typical of rural societies is the only thing to have been swept away by modernity. It is precisely our relationship with food—summed up in the verb "to eat"—that has been cut off from its traditional role in the history of humanity. The link between us and the world that surrounds us, the one that holds together the complex system of our existence, has snapped. This is why, irrespective of their relative degrees of modernity and wealth, traditional societies that, albeit unconsciously, have always lived profitably in a holistic way still have a lot to teach us.

Today eating generates uncertainty, anxiety, and fear. By barring nature from the human sphere, we have ultimately excluded food as well. We have forgotten the importance of an action that we perform at least three times a day, every day. The production and processing of food has left our kitchens to be taken over by third parties. Now that we have mislaid the secrets of food, we have to buy them back with money, just as we buy everything else we need—or think we need.

Food today is more a product to be sold than to be eaten. Reducing our relationship with what we eat almost exclusively to a series of market operations is both the cause and effect of a system that has removed value from food and meaning from our lives. It's a system that has turned the meaning of the verb "to eat" from active to passive for many inhabitants of the earth. Food has become as ambiguous as Dr. Jekyll and Mr. Hyde; given its complexity, it possesses "split personalities" in the way it appears and the way it is perceived. And split personalities are a sign of its unsustainability.

The Right to Pleasure

The idea for Terra Madre was developed by the Slow Food association, founded in 1989 and now boasting almost 100,000 members worldwide. Right from the outset, the association described itself as "a movement for the protection of and the right to pleasure," and also added the tagline as a subtitle to its manifesto. Twenty years ago, the description sounded to many like a fun provocation, almost like a joke. And that's more or less how the movement was received by the international press: especially in view of the way its name pilloried the rampant phenomenon of fast food, it was seen as something of an oddity.

A lot of water has passed under the bridge since then: through its initiatives Slow Food has managed to safeguard and protect small fragments of biodiversity and tradition, to educate many people about taste and eating, to found the University of Gastronomic Sciences, and to launch the Terra Madre network, to mention only a few of its achievements. Alas, on the pleasure front at least, the all-around work we have done for the cause of sustainability and better food has failed to achieve the results we were hoping for. Because it is held to be incongruous with commitment and "serious business," pleasure remains a taboo and continues to be denied.

The pleasure of food, like all other pleasures, is physiological, so there should, in theory, be nothing inherently bad about it. Instead, it is pointed at as being sinful and objectionable, and thus provokes holier-than-thou attitudes, disapproval from "fellow travelers," warnings from the health conscious, and accusations of shallowness. This viewpoint, still a thorny subject, dates from the Middle Ages, when ideas of "regimen," measure, and diet were overridden by ideologies and ways of thinking founded on the notion of "excess": excess in abundance, or overindulgence, but also excess in privation, or self-restraint. On the one hand, power manifested itself in the most sumptuous luxury; on the other, privation was seen as a sign of perfection and sanctity. The subtlest pleasure—along with sex, to which it is closely linked (both are necessary for the survival of the species)—was eating. And to enjoy that pleasure properly, the trick was to strike a balance between overindulgence and self-restraint.

It's still the same today. In consumer society, power is measured above all by economic wealth and associated in direct proportion to money. Wealth exhibits itself in opulence, the ostentation of pleasure, and the power to buy whatever one wishes. Even someone who isn't rich shows off whenever he or she can. It is arguably the only way left for many people to assert their existence in the wasteland of consumer society. But a poor person can in no way display power and is hence forced to renounce pleasure. In many spheres—religion, political ideology, ecology—the relationship between pleasure and virtue may be experienced by some as an incurable contradiction. If pleasure is excess and only the wealthy can afford it, it is regarded as incompatible with the pursuit of sanctity, political commitment, and the protection of the environment. The sentence is final: no pleasure!

It is thus normal for anyone not acquainted with Slow Food, with Terra Madre and the principles of a new gastronomic science, to exile "a movement for the protection of and the right to pleasure" into the enclaves of the bon vivants, the wealthy, the people who can afford luxury (luxury products generally being "niche" products), the superficial, and the uncommitted.

These prejudices make it hard for Slow Food to communicate ideas and initiatives; it's the price we have to pay. The most serious problem arises when the negation of pleasure—despite the paradoxical corollary that consumer society actually craves excess—ultimately becomes a perfect assist for the global agro-industry system. The negation of pleasure is a negation of the capacity of our senses, hence our power to understand and choose. The growing, processing, and distribution of our food are thus delegated to powerful third parties, and we are forced to accept the situation without even knowing what the consequences of certain methods of production will be. All this favors standardization, the result of an attempt to industrialize something—agriculture—which, by its very nature, cannot be industrial. This is the result of a system that, on the one hand, has succeeded in not feeding almost a billion hungry people and, on the other, in creating new pandemics, such as obesity and diabetes, which affect as many as two billion people around the world.

If we consider the resulting high health and social costs, we can only conclude that there is hardly anything virtuous about this development model. It is equally a paradox that, on the one hand, thanks to their local economies and quality produce, a lot of poor farmers fill the bellies of a privileged wealthy few, whereas, on the other hand, a few millionaire industrial farmers practice low-quality, intensive, monocultural agriculture to feed the poverty-stricken masses and the less well-off. None of this makes sense in light of the ongoing economic and environmental crises. We need to develop new forms of behavior, if we are to leave behind the prejudice that pleasure cannot go hand in hand with commitment. Not only are the two values—pleasure and commitment—

compatible, they combine to form a new and more creative way of seeing things.

Pleasure is democratic. It has to be because the pursuit of it, if responsible, allows us to interpret reality with heightened senses and sharpened intellect. Pleasure is democratic because it makes us want to become active players again, even if this only means performing small acts to improve our daily lives. The pleasure of eating is potentially the most immediate and accessible pleasure for all of us. And eating pleasurably may be a disruptive political act. Pleasure is not elitist; it is a right that needs to be protected, promoted, and enjoyed by all.

Tradition and Innovation

If we are to talk sensibly and without preconception about food, if we want to correct the global agri-food system somehow, we first have to debunk another cliché. Just as the economies of local communities are regarded as marginal and the pursuit of pleasurable eating perceived as elitist, so traditions—meaning ancient knowledge and simpler lifestyles—are cloaked in entrenched prejudice and branded as nostalgic and removed from reality. All this entails writing off centuries of popular culture as somehow "dated," with the result that most of the knowledge—or at least the origins thereof—of food communities is not even taken into consideration.

It is a paradox that the majority of people acknowledge the superiority (which they may regard as the privilege of an elite) of many traditional artisanal food products, made locally with fresh, seasonal ingredients and also consumed locally, but then they fail to give the cultures and skills that created them their due. It's like saying, "Yes, this food is better, but it's not to be found in the real world. It only survives in small niches, so we might as well eat worse." I think it's wrong just to give in without looking for alternative paths to follow.

I am convinced that, precisely on account of their wisdom and skills, food communities will be leading players in the third industrial revolution. I don't intend to be provocative here: I simply wish to point out that, if the world is demanding clean energy, sustainable production methods, the implementation of reuse and recycling schemes, the reduction of waste, the increased durability of commodities, and fresh, healthy, quality food, the food communities are not only runners in the race but are indeed already well ahead of the field; thanks to their techniques but, even more so, to the mentality behind them.

In most cases it would, of course, be impossible to export and replicate their methods—often based on very limited technologies—everywhere; they are the result of local adaptation, and it is at the local level that they work best. Nonetheless, it is fundamental to study their systemic nature—by which I mean their harmonization into a complex system—and understand the reasons behind it.

It is wrong to place traditional and popular wisdom a rung below academic science and privately funded research on the knowledge scale; the two branches possess equal dignity. Peasant know-how is the fruit of centuries of experience, and whether its practicality is scientifically demonstrated or demonstrable is immaterial. Not that I am arguing in favor of the primacy of this kind of learning, which I define as "slow knowledge." A dialogue has to be established in which prejudices are set aside, research is put at the service of slow knowledge, and research and science collaborate on an equal footing.

A frequent mistake is to view tradition as immobile and static, as a thing of the past. Even people who evoke, describe, and honor tradition often risk viewing it as a non-evolving thing, a singular curiosity that came to a standstill at some point in the past. This is a vision that ultimately cuts us off from our roots, depriving us of the memory of what we were and the history of our people.

This is something food communities know full well. For them, tradition is not a monotonous repetition of gestures and rites and products. They are open to novelty, to any idea that, following tradition, will enable them to progress; they are acquainted with the oft-cited concept of tradition as "a successful innovation" and put it into practice. They do not abandon the old for the new; on the contrary, they add the new to the complex system that has forged their identity. They know where they come from and are pretty clear about where they are going to.

We don't have to decide which is better: tradition or progress, past or future. But we do have to avoid generalizing, simplifying, and pitting the two concepts against each other.

Food communities are for the continuity of tradition, and they guard their memory precisely because it guarantees their identity in an increasingly standardized world. But they are also aware that it would be a serious mistake not to exploit the resources that globalization and technology make available. All they ask is to do this responsibly and sensibly.

True Unsustainability: The Problem of Paradox

We are all in some way to blame for the crises that have hit our planet. Our behavior, whether we like to admit it or not, has never been wholly virtuous. In many cases, we are unable to remain totally extraneous to the system, and we are forced not only to sustain it, but also to support it. In the so-called developed world, consumer society permeates our lives so thoroughly that we just can't opt out of it. Not that we should feel mortified or atone for our sins through extreme sacrifice (negative excess); what we should and can do, though, is to set into motion slow but undeniably positive processes. To do this we have to change our mentality. We have to learn to open our minds to the "not-exact" and the "not-fully-explained," to the good and the beautiful—concepts that cannot always be universally codified. Maybe we need to think a little more systemically, casting anxiety and uncertainty aside. For anxiety and uncertainty are fruits, more than anything else, of a development model that seeks to control and pigeonhole what cannot be controlled and pigeonholed.

It's not paradoxes that we have to fear but a lack of commitment in overriding them; that, for sure, would be an unsustainable way to behave. Perpetrators of food paradoxes accuse people who think differently from them of contradicting themselves. This should not concern us: if we consider the whole of which they are part, any human action or way of thinking may appear to the inveterate pigeonholer as a bundle of contradictions. But, as we have seen, it is possible for apparently contradictory things not to be separated and to coexist, provided you switch the mental planes on which you arrange them. "I don't generally trust people without contradictions," a Spanish friend once told me. How right he was! If you consider the plurality of identities and the diversity that surround us, you'll realize that it is practically impossible not to contradict yourself every once in a while—and there's nothing wrong with that. But coherence is a different matter; it means being clear about the fact that you don't wish to jeopardize the health of the planet. It means giving communities the opportunity to eat freely what their members choose to grow. It is a guarantee that food is and will remain the best form of peace diplomacy available to us today, as well as the greatest source of pleasure and happiness. There are a million and perhaps even more ways of achieving all this; as many as there are vegetable varieties and animal breeds, multiplied by the number of ways of cooking and processing them, and multiplied in turn by the number of ways of eating them. We have a virtually never-ending and, luckily, noncontrollable list of solutions at our disposal to find the right ones to our problems and to our "crises."

We have to override paradoxes and have no fear of contradictions. We have to leave unsustainable ways behind and restore the proper meaning to the verb "to eat."

The Foundations of a New Gastronomy

Pleasure in itself is not a deviant form of behavior. What is deviant and unsustainable—insofar as it generates unsustainable behavior—is the belief that pleasure is something that cannot be disconnected from excess, in both a positive and negative sense. But it is excess that needs to be declined, not pleasure in itself.

In the form of Pantagruelian abundance, excess may confound our ideas about our own limitations; in the form of privation, it may muddle us up as to our potential. Pleasure as such is a physiological factor that cannot be denied; indeed, to feel, recognize, and search for it is a constructive thing to do. But it is also necessary for us to be aware of our limitations and potential and to refuse all forms of excess. This is not so much a matter of finding the "right means" as of using common sense. It is what Bartolomeo Sacchi calls *honesta voluptate*, which we might define today as "sober pleasure."[1]

It may seem easy to talk of sobriety in times of crisis, and obvious to see it as the immediate consequence of the financial and economic situation. But sober pleasure is neither moderation for the sake of it, nor forced temperance, nor "scrimping and saving." I want to avoid adopting a mortifying tone, precisely because I would end up sounding punitive. We claim the right to pleasure with the same responsibility as food communities and the food community we belong to. Pleasure is a gift of nature. Contrasting it with commitment and ethics doesn't help us to understand its beauty. Nor does it help us to understand that, precisely because it is natural, pleasure should be harmonized with nature, so that anyone can be entitled to it, in the present and in the future.

Sobriety means avoiding waste, taking advantage of renewable

---

1. Bartolomeo Sacchi, nicknamed Platina (born Piadena, 1421; died Rome, 1481). A fifteenth-century humanist famous above all for his brief treatise on gastronomy, *De honesta voluptate et valetudine* (Sober Pleasure and Good Health), Platina transcribed into Latin all the recipes of Maestro Martino da Como, the most celebrated cook of the time. Platina was the first person in history to analyze gastronomy, diet, the value of local food, and the importance of regular exercise—all subjects that are still topical today. The treatise is inspiring above all for its idea of *honesta voluptate*, a form of conduct in which pleasure is not denied but exercised soberly and responsibly.

energies as much as possible, learning the lessons that all previous generations at all latitudes have handed down to us. These lessons serve us in good stead when we have to face periods of difficulty, turning the privation caused by poverty into an element of conviviality, into an excuse for inventing a tasty new recipe—or celebrating abundance and sharing it with others. Sobriety often coincides with simplicity, a quality closely linked to the common sense I spoke of above. It is a way of reacting, each in one's own context, of exploiting the exploitable, but without jeopardizing it. This is what the food communities lay claim to, and this is what could represent a fuller life for all of us—a life in which we are the masters of the situation.

Not that it's my intention to paint an idyllic picture of a bucolic and utopian dream. I do realize that I am opening myself to more criticism here, so I don't want to forget or sidestep the objective difficulties involved in my argument. We all know that rural societies have never been places in which it was easy to live. In many parts of the world, the harshness of their kind of life was, and still is, terrible. From their culture I want to salvage the features that may be innovative and positive; features far removed from our experience as *Homo consumens*, but which take us back to forms of behavior comparable to that of coproducers, to respect for certain human values and for the earth: to the community context, to the spirit of local adaptation,[2] to learning about the rhythms of the seasons, and to in-depth knowledge of food.

This, in a nutshell, is the new gastronomy I am advocating, one that marks radical change in our cultural, economic, and social

---

2. Wendell Berry, *Life Is a Miracle: An Essay Against Modern Superstition* (Berkeley, California: Counterpoint, 2001).

approach to food. New gastronomy is a science, but also an attitude. To explain exactly what I mean, let me quote Jean-Anthelme Brillat-Savarin's definition of gastronomy in his *Physiology of Taste* (1825): "Gastronomy is a scientific definition of all that relates to man as a feeding animal." According to Brillat-Savarin, gastronomic science is complex and comprises many different disciplines: natural history, physics, cooking, commerce, political economics, and so on. Even the "actors" of gastronomic science, argues Brillat-Savarin, are not only "savants," but also "cultivators, vine dressers, fishermen, huntsmen, and the immense family of cooks, whatever title or qualification they bear to the preparation of food."[3]

If we accept this definition, the upshot is that gastronomy today has to be redefined—in light of the evident and rapid transformations of society in the postmodern era—as a complex, interdisciplinary science that studies food and everything inherent therein. The subjects listed by Brillat-Savarin are still valid, but today we might also add anthropology, genetics, animal husbandry, agronomy, sociology, medicine, history, and also ecology.

Food should be studied as culture, as a raw material and a commodity, as an artisanal, industrial, or culinary product—and as the act of eating itself.

Gastronomy is thus a multidisciplinary science, one that involves and intersects with all knowledge of food as a material element, from its cultivation (or breeding) to its consumption, and as a cultural element, processed according to tradition or otherwise, recounted or analyzed in a more or less scientific way.

---

3. Jean-Anthelme Brillat-Savarin, *The Physiology of Taste*, trans. Fayette Robinson (Adelaide, Australia: University of Adelaide, 2007).

This approach obviously encompasses all the elements that food communities are endowed with: know-how, local products, traditions, food-related rites, attitudes, and even ways of thinking. Gastronomy respects all these features and in no way regards them as backward. Unprejudiced as it is, it adapts to their holism, to the way of understanding nutrition and life of producers and coproducers alike. Gastronomy is a science that, by definition, places no barriers between disciplines and admits "hidden" connections, even if they are often apparently unfathomable.

Realizing that sustainability cannot be ignored, the new gastronomy promotes, in turn, a new agriculture. The new gastronomy achieves sober pleasure thanks to a complete perception of reality through trained senses, comprehensive knowledge of food, and the capacity to choose what is best for the individual and the community. The same sobriety accompanies research and the claim to the right to pleasure; it lays a foundation for a new sustainability and a new humanism.

FOUR

the value and price of food

The triumph of consumerism has seen the triumph of another prejudice-cum-cliché: the idea that the price and value of food has to be low—as low as possible, in fact.

It's natural, in a market, for us to opt for the product that costs least. But we should do so when quality is equal, or at least when we have the opportunity of choosing a standard of quality suitable for our needs. This is no longer possible in the case of food; it has to be cheap, period. Vegetables or pasta only have to go up by a few cents and the papers spew indignant reactions. Yet people don't protest the same way if their bank account or telephone bills cost more, if a professional fleeces them for his services, or if a television-repair call costs the equivalent of a dinner for two at a restaurant.

But food's a different matter; it isn't to be meddled with. The widely held opinion is: "With a great deal of effort we managed to beat hunger years ago. We are a rich, opulent society; food has to be available everywhere and, if possible, cost a trifle. If it's expensive, let's leave it to gluttons and guzzlers with plenty of money to spend." This is what comes of having transformed food into a consumer commodity, stripping it of all its spiritual, cultural, and material values: the system built around it or of which it is part has replaced value with price. Money has supplanted other values to become the secret of happiness.

Food is thus no longer produced to be eaten, but to be sold. Price becomes the principal, if not the only, choice criterion. In the global agro-industry food system, foodstuffs have become commodities just like all the others—no more, no less; just like oil, timber, or other tradable goods whose prices are established by international stock exchanges. Grain, corn, coffee, and cocoa are commodities like metals or energy, hence subject to the laws of supply and demand, distributed on the market without differentiations in quality and without a care about who produces them.

Subjecting food to these laws leads to a standardization of food production that tends to reduce biodiversity and increase "eco-unfriendly" monocultures. And it also causes a huge amount of injustice. Especially in the South of the world, and often on account of their colonialist or neocolonialist heritage, whole countries have become specialized in given agricultural products and promptly suffer huge upheavals when their prices plummet.

Mostly in countries that are experiencing rapid urbanization, the fact that food is becoming something to buy and not to produce is creating poverty, hence hunger and malnutrition. A peasant farmer in a poor country who decides to abandon the hard life of the countryside for a move to the city stops producing the meager amount of food that allowed his family to get by, albeit in poor conditions. But if he doesn't find a job with a decent wage in the city, he won't be able to buy enough food for himself and his family. In a short space of time, he will descend from poverty to nothing—to hunger and downright squalor.

Total commodification is the price we pay for the degeneration in the value of food in both the North and the South of the world. Yet practices exist that may not be viable in monetary terms—think of home jam making, which certainly costs less than buying jam in a store—but which enables us to earn value in terms of conviviality, personal gratification, community service, environmental protection, and, in a word, well-being.

We have perpetrated the most appalling disasters, all for the sake of Mammon. Destitute of authenticity, food ultimately eats us. Deprived of cultural, social, and environmental values, it stops being an object of attention, care, and pride—as a true resource should be—morphing into a monster that devastates the countryside socially and ecologically, causing injustice everywhere. We may think we can nonchalantly throw away food, but we can't.

Food Is Eating the Environment

Industrialization, as intense in the agri-food sector as it is in others, relegates quality to the back seat. The drivers are quantity, productivity, standardization, and homogenization. Nature, characterized by complexity, indeterminateness, diversity, and multifunctionality, is something else.

Industrial agriculture (what an oxymoron!), the industrial processing of food, the distribution over five continents of foodstuffs that could be cultivated *in loco*, low prices, and the laws of the free market—all these factors have combined to make the food sector one of the most unsustainable spheres of human activity.

Over the last hundred years, biodiversity has disappeared at an alarming rate: the need for vast monocultures to supply industry with large amounts of cheap food has limited people's choices to the few varieties suited to this production model—to the detriment of others. As a result, in the United States alone—the world leader in industrial agriculture—80.6 percent of all tomato varieties became extinct between 1903 and 1983, as did 92.8 percent of salad varieties, 86.2 percent of apple varieties, 90.8 percent of corn varieties, and 96.1 percent of sweet corn varieties. Of the 5,000 existing potato varieties, only 4 constitute the majority of those cultivated for commercial purposes in the United States. Only 2 varieties account for 96 percent of all cultivated American peas, and 6 varieties for 71 percent of the total of all cultivated corn. These have been the results wherever the industrialization of food has had the upper hand: a triumph for standardization and homogenization, and a serious peril for two of the cornerstones of life on earth—biological diversity and the ability of species to adapt.

The damage that has been done has achieved biblical proportions. In just one century, we have allowed the fruits of thousands of years of evolution to vanish. Luckily, many countries that have yet to experience the agro-industry boom still enjoy a decent level of biodiversity. But if their ambition is to emulate the example of the West, the disaster will become universal. Unfortunately, we are already seeing worrying signals of how, in countries such as Mexico, India, Brazil, and China—among those boasting the most edible vegetable varieties and animal breeds, but also with galloping industrial growth rates—the phenomenon of the destruction of biodiversity is repeating itself with unprecedented intensity.

Furthermore, even the land is being "eaten" by food on account of production being carried out on an industrial scale. Over the last few years, the use of chemical fertilizers and pesticides has increased exponentially. The same amount of synthetic chemicals has been applied to the world's soils and introduced into its natural systems in the last ten years as was used in the whole of the preceding century. The products in question are, of course, foreign to the natural cycle, and it is no secret that they will jeopardize soil fertility in the long term. The soil is a living thing, and we are murdering it. Industrial agriculture has embraced the idea of farming without farmers, but at this rate one day we'll be forced to farm without land.

Damage to the environment caused by the industrial global food system is so widespread and severe that the problem is now the first item on the ecological agenda. News about what is going on has leaked from the "alternative" world of organics and environmentalism and now entered the public domain. It's no longer possible to deny the facts.

The countryside used to be an oasis for town dwellers keen to escape from pollution. Today many areas of the planet have become dangerous for our health, especially places where fertilizer is spread and pesticides are sprayed. Agriculture used to be—ought to be—an alliance between man and nature, but it has gradually become a war. It's no coincidence that the technologies used to produce pesticides all originate in the armaments industry. Industrial agriculture is de facto a declaration of war on the earth.

To date, environmental devastation has never been calculated as an item in food economics, even though it does represent an increasingly onerous cost.

I believe that we pay a low market price for food, but that we also pay a high—and hidden—price, not only in economic terms, but also in terms of the earth's capacity to produce food in the future, and in terms of the quality of our own life and health and of those of future generations, to whom we cannot deny the sacred right to enjoy well-being and happiness. The low cost of food not only devalues food itself, but also hides all the evil we are doing to the earth.

Sooner or later, someone will have to pay for all this, and ultimately it will be "consumers," even if they are convinced they are getting a bargain when they spend small sums of money on eating.

Food Is Eating Farmers

Besides seriously damaging the environment and nature, the industrialization of food has also caused social upheaval, in the countryside, in towns, and in modern megalopolises.

The depopulation of the countryside, a process that began in Europe and the United States as early as the Industrial Revolution of the nineteenth century, but escalated sharply in the years following World War II, has spread to all the countries that have industrialized their agriculture. Fifty years ago, half the population of the Western world worked in agriculture in some capacity: today the number has dropped to between 2 and 7 percent. The global South has still to reach these percentages (the number of people living in towns and cities overtook that of people living in rural areas just two years ago), yet great metropolises such as Mexico City, São Paulo, Mumbai, Lagos, and Beijing are bursting at the seams with desperate former farmers. Dazzled by vague hopes of fortune, worn out by the tough life of the countryside, strangled in the vise of agro-industry, which is profitable only on a large scale, they sell their land in the hope of building a new life for themselves in the city; but the vast majority of them simply swell the mass of poor suburban slum dwellers.

The growth of the world's megalopolises is staggering and would appear to be unstoppable. In 1950, there were 86 urban areas in the world with over a million inhabitants, whereas today there are 400, and in 2015 the figure is expected to rise to 550. The aggregate urban population of China, India, and Brazil today almost equals that of the whole of Europe and North America put together. The phenomenon is fueled largely by higher urban-population growth rates, but the depopulation of the countryside is also a contributing factor. All this would have a positive side if it were able to secure better living conditions at least for the farmers who stay in the country to ply their trade. Alas, workers in large-scale industrial agriculture in the North of the world are no longer masters of their own labor. Farmers, transformed into piece workers at the service of the food multinationals, no longer own their own livestock or the seeds of the crops they grow. Industrial-agriculture systems have taken everything away from them, leaving them to get by with less and less fertile land, livestock raised in assembly-line fashion—hence in precarious health—and vast monocultures of soy, corn, or canola that are starting to cause huge problems.

Since large-scale agribusiness has very low profit margins, the economic advantages of farmers participating in it are fewer and fewer. All it takes to bankrupt large family farms with a long history behind them is a variation in price or a spell of bad weather. The model whereby a few wealthy farmers produce large quantities of food—in some cases of inferior quality—for huge masses of people has proved to be an illusion. Such farmers do not earn enough; indeed they work at a constant loss, and their businesses only survive thanks to the large subsidies Western governments grant to prevent domestic agriculture from dying altogether.

Not that farmers in the South of the world are faring any better. On the contrary, for many of them the only alternative to going to live in the city is to commit suicide. Every year in India at least 20,000 peasant farmers take their own lives because they cannot afford to pay off the debts they incur when they buy seeds, fertilizers, and pesticides. This startling statistic reveals how, when applied to the small scale, industrial agriculture has truly devastating effects, even in terms of human lives. You would be excused for thinking you were reading a war bulletin.

The depopulation of the countryside and the wholesale impoverishment of farmers inevitably tears apart the social fabric of communities. Seen through the farmers' eyes, the face of industrial food is inhuman and harrowing. All industrial food does is discourage them from doing their work, from caring for the environment, and from producing good food for themselves and for their communities. This is another result of the cheap-price policy—food is eating farmers. And waste is consuming all of us.

When the price of food is low and its value decreased, it is almost natural to waste products without a thought. This is happening ever more dramatically, ever more unjustly with food. Recently the number of people suffering from hunger topped the one billion mark. Though countries have set themselves a common task of considerably reducing the number by 2015, it is still rising all the time.

This is why figures on waste leave us speechless. In Italy, according to research conducted in 2007 by Siticibo, an association that works in conjunction with the Italian "food bank," Banco Alimentare, we waste 4,000 tons of edible food every day; that is to say, 1.46 million tons a year.

In the United Kingdom, according to the Waste and Resources Action Programme (WRAP), they waste 6.7 million tons a year, about a third of the total available.[1] According to the U.S. Department of Agriculture (USDA), Americans waste 25.9 million tons a year, a quarter of all their food. But a study conducted by the University of Arizona in 2004 pushed the figure up still higher, showing that in some cases it can account for 50 percent of the total.[2] A curious and shocking figure is that of the rice wasted by Filipinos: 1.2 million tons *every day*, according to the Philippines National Food Authority.

How is it possible that the industrialized countries have managed to reach this level of contempt and neglect for food? As we have seen, the laws of consumerism have a lot to answer for. One of their biggest responsibilities is their continuous pursuit of the new and their apparently qualm-free abandonment of the old. Packaging, product durability, increasingly large portions, and ready-made meals—everything in the global-industrial food system works against storage and saving. To understand what waste really means, take a glimpse behind the scenes at a supermarket. Up front, the place looks like the land of plenty, but, backstage, it's more like a garbage dump: packaging, products just past their sell-by date,

---

1. See http://wrap.s3.amazonaws.com/thefoodwewaste.pdf.
2. Timothy W. Jones, "The Garbage Project" at the University of Arizona, http://uanews.org/node/10448.

fruits and vegetables no longer biologically or aesthetically presentable on the shelves. All unsold goods that get thrown away.

But never mind large-scale distribution, what about our own refrigerators? They were invented to store food, to store leftovers for reuse. Today, full as they are with jars of jam showing the first signs of mold, old cheese rinds, half-finished cartons, they are more like the last step before the garbage bin. We throw too much away. Home freezers are the most glaring manifestation of our atavistic memory of hunger: our craving not to find ourselves foodless leads us to bury meat in ice—and once it's in there, it may not resurface for years.

The art of storing food used to be a matter of life and death. Now that we have handed the job over to technology, we have become frenetic producers of refuse. The wise old school of home economics has become an optional extra in our "disposable" culture. So it's no coincidence that the subject doesn't get taught any more. Since it used to be taught mainly in girls' schools, they decided it was simpler to get rid of it on the grounds of sexual discrimination than teach it to boys.

You don't need to go back that far to remember the period in which our society, caught up in a whirl of well-being, elevated the disposable object to symbol status. In 1994, in his seminal essay "Un mondo usa e getta" (A Disposable World), Guido Viale, one of Italy's leading experts on waste, cited one of Italo Calvino's "invisible cities" to exemplify the garbage-dump direction modernity is taking: "The city of Leonia refashions itself every day.... On the sidewalks, encased in spotless plastic bags, the remains of yesterday's Leonia await the garbage truck.... It is not so much by the things that each day are manufactured, sold, bought that you can measure Leonia's opulence, but rather by the things that each day are thrown out to make room for the new. So you begin to wonder if Leonia's true passion is really, as they say, the enjoyment of new and different things, and not, instead, the joy of expelling, discarding, cleansing itself of a recurrent impurity."[3] Ultimately an avalanche of its own waste will bury Leonia, "canceling every trace of the metropolis always dressed in new clothes."

To think that, at the preproduction phase, we could do a lot to produce less waste simply by returning to healthy domestic economics in collaboration with producers, distributors, and consumers (not necessarily in that order).

It is Leonia-style mania that has caused us to forget popular wisdom, a wisdom that teaches us how to produce food better, release less carbon dioxide into the atmosphere, respect biodiversity, and avoid waste. We have to learn how to do our shopping again—which means becoming coproducers: avoid buying more than we need and refuse to bow down to the logic of preprepared portions and ready meals.

Imagine the amount of paper or plastic—the polystyrene and film used to wrap too many apples for a single person and too few for a family, or all the cardboard and plastic that yogurt is packaged in—we could dispose of during our wait in the supermarket's checkout line. In some countries, supermarkets have begun to sell loose products, which customers transfer into containers they bring from home. These are virtuous forms of behavior that, hopefully, will induce even producers to see the uselessness of much of their packaging and reinvent it in more eco-compatible

---

3. Italo Calvino, *Invisible Cities*, trans. W. Weaver (New York: Harcourt Brace Jovanovich, 1974).

terms. This would generate huge collective savings and less refuse production.

Italy's housewives used to be experts in home economics because they had to feed families with the paltry resources available to them. They were so good at this that they managed to create many of our traditional gastronomic jewels from leftovers. There's an Italian proverb, "*Del maiale non si butta niente*," which means that, after slaughtering, "No part of the pig is thrown away." The utilization of leftovers was part of a way of approaching life in which elders were people who were listened to, not nuisances confined to a rest home. Old, used-up belongings were remembered for the role they had played in the family, and it was always deemed a pity to get rid of them. You even postponed replacing them; waste was a crime. Today leftovers are no longer turned into delicacies by skilled hands; as in Leonia, they are thrown away or hidden every day.

There is far too much waste food around. There are mountains of it, and it's going to bury us. Food will end up devouring us all.

Donation as Waste Prevention

In 2003, I had the honor and good fortune of being the guest of the Trappist monks at the Abbaye Notre-Dame de Saint-Remy in Rochefort in Belgium, brewers of Rochefort, one of the best beers in the world. I felt immensely privileged to be able to enter the abbey, see the production lines, and interview the monks. At one point on my visit, standing beside two huge copper vats, one of them told me, "In the first one we make the beer, with the other we do nothing. It only comes into action to replace the first one, if it stops working."

"Why not use both of them?" I asked. "You make one of the world's finest beers, and it's a best seller. You could double your output and your profits."

The monk replied with a simple, straightforward explanation of the monastery's productive philosophy. "We don't need to double our output. You see, each year we have a meeting to figure out how much beer we have to produce to meet the needs of our community. We calculate our plant and production costs, the cost of our labor and the labor of the laypersons who help us, plus how much money we need to keep the monastery and its facilities running. We are pretty thorough about all this. Which is why we are able to work out the exact amount of beer we have to make in a year—the amount that will ensure us the right income. We then add on an extra amount to generate money for charity. Our total output is thus pre-calculated on the basis of necessity and of donation. We don't need to produce any more than that."

I was taken aback. This is what I call having a sense of proportion: producing for real needs without going over the top, and producing without greed. On reflection, the most important thing the Rochefort monk told me was the amount of surplus beer produced for donation. Here we have a representation of an "economy of freewill giving" that has been swallowed up by the vortex of consumerism, in which profit is the true religion and everything has to generate an economic return.

Freewill giving is still a fundamental value in food communities and has always been part and parcel of peasant civilization. It was, and still is, a common practice in farming communities—

though no longer elsewhere, alas—for families to help each other. For example, the first family to complete the harvest will help its neighbors to complete theirs, lending its machinery if necessary. Families also give seeds to others who have lost their own. In the past, it was even a habit in many Italian farmhouses to put an extra meal on the table, in case a traveler happened to knock at the door. Of course, many forms of freewill giving may appear archaic and old-fashioned, but they don't necessarily have to be repeated in exactly the same way as in the past.

Allocating a production quota for donation has a very interesting economic function. I believe, for example, that donation is the best way to prevent waste. If I plan ahead to give something away—be it goods, food, time, or labor—I fit it into the economic mechanism as a sort of shock absorber. Its effect is to monitor the economic, ecological, and existential sustainability of production activities and avoid wasting any surplus. I'm not referring here to the niggardly "humanitarian aid" allocated to countries in the so-called Third World. This has ultimately brought entire economies in Africa and Asia to their knees, devastating local markets and forcing people to abandon the cultivation of native products. This type of donation is a cover for self-interest: the need to dispose of the production surpluses of the North of the world, bloated by grants and subsidies. No, the donation I'm speaking of has to be made within the local economy through humane, natural forms of production. Building a donation into the production budget provides a safety net if production doesn't go as planned, and, if it does, it goes to help others and, most important of all, avoids waste. Most of the 4,000 tons of edible food we throw away in Italy every day could be donated to people who are in dire need of it (according to a study published by the Banco Alimentare "food bank" on October 8, 2009, 3 million Italians live below the poverty line and are undernourished). Consider these figures and you can understand why making freewill giving a part of economic discourse may be of great environmental, social, and economic value. This value is denied for the sake of production and consumption, yet it characterizes a product in a different way from price alone and restores to food part of its true meaning.

Speed, Abundance, and Induced Needs

According to FAO data, enough food is produced in the world for 12 billion people, but the population of the world today is just under 7 billion.[4] Against all logic, we are seeing a constant thrust to increase production. The multinationals are always on the lookout for new ways of boosting quantity, and appeals from the people formally responsible for addressing the hunger problem are invariably along the lines of "We need more food!"

When it comes to crisis solving, governments always advise us to consume more. They just don't get the fact that it is our crazy, muddled, wasteful consumerism that got us into this position in the first place.

But there are other reasons why we allow ourselves to be eaten by food. One is the speed of our society which, in a whirl of production and consumer frenzy, causes us to lose all perception of the

---

4. United Nations General Assembly, *Promotion and Protection of All Human Rights, Civil, Political, Economic, Social and Cultural Rights, Including the Right to Development*, report of Jean Ziegler, special reporter on the right to food (January 10, 2008), A/HRC/7/5.

reality we are living in. It forces us, as consumers, to chase promises that inevitably remain unfulfilled, and thereby exhausts us. Speed deprives food of value and either reduces it to fuel without quality or raises it to a status symbol of elite consumption, with strong connotations of added values—no middle ground exists. Yet common sense tells us that it is precisely halfway between these two extremes that the true value of food lies. Speed prevents us from seeing our limits, and thus forces us to go beyond them. We only realize we have bitten off more than we can chew (so to speak) when we make mistakes, many of them irreversible. This is why I rate slowness as an important value: "Slow" means learning to know your limits, savoring and distinguishing differences, not underestimating durability, by which I mean not seeing everything as old and dated. Slow movers have time to look over their shoulders and use their memory without erasing it or having others preserve it for them.

Another great illusion of consumer society is its sense of living amid abundance: "much" and "a lot" don't necessarily mean "quality," and, above all, they have nothing to do with humanity. Resources sufficient for everybody are out there, but the problem is that we don't know how to use them, thus creating absurd imbalances both outside and inside ourselves. Abundance makes us waste food—though, at the same time, making us demand that it be cheap—and has made us prisoners of a system in which the value of food is no longer recognized.

As we all know, it is in the nature of consumerism to induce needs, making us buy what we don't need and selling us false promises and false values. Food-industry advertising is scandalous in the way it spreads falsehoods, devalues the sacredness of food, misrepresents pleasure as excess, and thus eats into our minds. TV commercials sell our brains to food producers. The way they expropriate the brains of children in particular is criminal. In Europe, children sit in front of the television an average of three hours a day, and are bombarded by advertising in which food becomes everything except what it really is. In Italian TV commercials, for example, we see a cow coming into a house in the morning to be milked and a grandmother shooing out the family because she says she has no food in the house, then sitting down to "savor" a prepackaged, ready-to-serve meal on her own. Really! It's grandmothers who should be teaching us about food and taste. Even more than mothers, they ought to be handing down the joy of food and their knowledge of it.

By telling us all about needs we never even dreamed of having, advertising constitutes the most obvious proof of the fact that we are no longer masters of our food and our future.

Eating Well Isn't Expensive

It's true, eating well is not expensive. If, in times of crisis, the only alternative is to eat at a fast-food joint or buy low-quality food at a discount store, maybe this means that we have problems that are more serious and ingrained than the crisis itself. Attempting to fix the family budget by serving poor food that isn't good for you in the long run is no solution.

There are other ways of addressing our daily food: think good meat, good fish, and good vegetables. Everyday food and even the occasional meal out can actually cost very little. And food can be of good quality and good for us in terms both of personal health and of public health.

We have to forget the old prejudice whereby good food is only for the elite. Above all, we have to follow two simple rules, seeking quality outside the consumer system and rediscovering good domestic and gastronomic practices.

Rule one. To exit the system it is necessary to resort to alternative distribution channels. In every town and city there are markets in which you can buy better-quality produce directly from farmers at advantageous prices. It's fundamental to respect the seasonality of foodstuffs; in season, fruits and vegetables cost less.

To buy fruits and vegetables, you don't even have to bother to go to the market. In Italy, for example, Gruppi di acquisto solidale (GAS)—fair-trade shopping groups—are increasingly popular and widespread, while there are also producers' cooperatives that guarantee direct home deliveries.

Rule two. In order to follow good domestic and gastronomic practices, we need to recover aspects of our culture that we have forgotten. Take cuts of meat, for example; the loss of artisanal skills in abattoirs, now little more than large-scale disassembly lines, means that a lot of consumable meat now gets discarded. The less noble cuts are no longer in demand because we no longer have the will or the way to cook them. The consumer has become fillet obsessed. La Granda, an association based in the northwest Italian town of Cuneo, has been raising Piedmontese cattle for years. It sells virtually every part of the animals it butchers; not only to make sure that breeders earn as much as possible, but also to let us save too. They tell me they only throw away the horns and hooves of the animals, and that they cut and sell the forequarter (including the chuck, flank, brisket, and rib muscles) and process what Romans call the *quinto quarto*, or "fifth quarter," meaning the head, tail, insides, blood, and trotters. The association has thus come up with products such as hamburgers (at 1.15 euros each); galantines (boned, stuffed meat that is poached and served cold under aspic); alternative canned meat (made with the cheeks, trotters, tongue, and tail); and individual portions of broth, pasta sauce, and paté. A fifth of La Granda's products are ready-made meals, all made without preservatives and with excellent, tasty meat that, according to a doctoral thesis in medicine at the University of Turin, has higher nutritional value than normal veal and beef.

The association's most prized cuts cost much more, of course—and rightly so, given its livestock-raising techniques—over 20 euros a kilo, more than meat produced using less-sustainable methods. But a kilo of superior-quality bull's beef, bought directly from the breeders and butchered in such a way as to avoid all waste, only costs around 10 euros. Of course, since the meat of the bull is less tender than that of the cow, you have to know how to cook it; stewing or braising is recommended.

We find the same lack of gastronomic culture when it comes to fish. Suffice it to think of species that are caught and then returned to the sea because they have no market, because there is no demand for them. Everyone clamors for sea bream or bass because they haven't a clue how to cook other species. Sardines, anchovies, mackerel, and the like are good, tasty, and healthy, but a trifle more difficult to prepare. Yet such fish is very, very cheap.

Another thing we have lost the hang of is food storage. I remember that, where I come from, in the summertime people got together in their courtyards to make conserves in huge pans from freshly picked, perfectly ripe tomatoes. They then poured the sweet-scented mixture into sterilized glass jars, which they sealed hermetically. Nowadays in the winter we demand expensive and tasteless cherry tomatoes from God knows where. A jar of homemade *passata* costs much less.

All these good practices and skills—this baggage of popular creativity—have been virtually abandoned. To think that they could be translated into savings, into solid cash.

If we are no longer prepared to source and cook good local produce, grown just outside of town, seasonal, and probably costing relatively little, we can't complain if food is expensive.

Overcoming Uncertainty

For many people, the future of the world is increasingly uncertain. But, on reflection, this shouldn't frighten us: the future is uncertain because it is unforeseeable by definition. This, of course, may cause a certain amount of anxiety, but such anxiety is a fact of life. The system used to seem perfect. Born of science that emerged in the Age of Enlightenment, it isolated processes and extrapolated them from their context to understand them better. If a problem arose, the system itself sought to improve processes, to invent new, more efficient ones, or even to build machines capable of reproducing them. We believed we had understood how nature worked; we were convinced we could predict, make improvements ad infinitum, comprehend everything.

We deceived ourselves into thinking that we were total masters of our own lives. We believed we could use the earth as we wished, to gain benefits and, above all, profit. From my standpoint, which is that of someone concerned with gastronomy and ecology, I can't help but notice how the lack of a systemic, holistic approach embracing all the connections that exist among the living systems has caused very serious damage.

Reducing food cultivation, processing, and consumption to a series of mechanical procedures, as if they were disconnected one from the other, has caused us to mislay the deepest meanings of an act—eating—that is indispensable for our survival.

In reality, the world is not plunging into uncertainty. Uncertain are the people who used to pretend to be certain, but who now grope in the dark, attempting to solve problems using the very system that created the problems in the first place.

The Terra Madre food communities teach us that there's another way out. They teach us that we have to start again from food, to make it central once more. We have to go back to basics and learn to understand food, to cultivate it, to grow it, to choose it, and to eat it, without letting ourselves be eaten. Only in this way will we find a better way of living.

FIVE

food sovereignty

In a world in which food is consuming us, that food also controls us. Food that is industrial, standardized, global, and unnatural. Food that is unsustainable and pollutes the earth, from the fields to our stomachs. Food that generates crisis and uncertainty. The people in charge of food—the food sovereigns—are the people who make it: the people who run food companies and agribusiness multinationals and the large-scale distributors who fix prices without a care for the interests of farmers and coproducers.

If we are to start changing this state of affairs, and restoring our proper relationship to food, what we have got to do is forge an alliance, an agreement, between the people who produce food and the people who consume it. All those who place, or would like to place, food at the center of their lives need to give it the importance and value it deserves.

All those who play a primary role in the creation of food must become "sovereign" again. They must be able to decide what to sow in the fields, which animals to keep in barns and which to put out to pasture, which varieties, breeds, and techniques to use. They have to see that their cultural traditions and local environment are protected, that what they produce is at the service of a living gastronomy—mirroring the history and habits of their people—and that the natural cycles and regenerative ability of the earth are respected. That is what I mean when I use the term "food sovereignty."

(Re)conquering Civilization

I support the fundamental principle of national, regional, and community food sovereignty. All local, national, and regional entities and communities have the inherent right and obligation to protect, sustain, and support all the conditions necessary to encourage the production of sufficient healthy food in a way that conserves the land, water, and ecological integrity of local areas, respects and supports producers' livelihoods, and is accessible to all. No international body or corporation has the right to alter this priority. Neither does any international body have the right to require that a nation accept imports against its will, for any reason.[1]

It is a paradox of our times that this principle, this right, still has to be asserted and that we have to launch constant appeals for it to be granted to people. You would think it would be the most normal thing in the world: "I own a field and I use it to cultivate what grows best, namely, local varieties, and I do so for myself and for my family first and foremost, after which I can sell any surplus I produce because there are people in my area and in the environs who appreciate my work and the care I take in growing food, and they pay me a fair price for it." But no, if we construe this long concatenation of clauses, we discover that the simple affirmations they contain are not so obvious after all; that verbal predicates such as "cultivate," "sell," "admire," and "pay" have become a trap for farmers and for

---

1. International Commission on the Future of Food and Agriculture, "Manifesto on the Future of Food," 2006, p. 12, http://www.arsia.toscana.it/petizione/documents/cibo/cibo_ing.pdf. I have chosen to use the definition of "food sovereignty," the most cogent in my opinion, proposed by the "Manifesto on the Future of Food" drawn up by the International Commission on the Future of Food, of which I am a member.

eaters. The middlemen in food's field-to-table journey have twisted the meanings of these verbs. They have reshaped them into a function of profit, so that we have to demand things that ought to be normal, things that ought to be ours by right.

Take the concept of "organics." If you think about it carefully, organics is something of a contradiction in terms. Why should we have to certify and label as natural stuff that grows without additives, in fertile, clean soil? It should be the norm, because that's how nature works. Absurdly, natural produce has become an exception that has to be certified; the rest, ruined by all sorts of artifice and injections of foreign substances into the natural cycle, has been transformed into "normal" food.

The conquest of food sovereignty, a fundamental principle of human rights, will ensure the promotion and dissemination of the wisdom of food communities. Food sovereignty will be their banner and the banner of the rights, devices, techniques, and cultural directions that encompass the whole arc of complex, interdisciplinary gastronomy. This system of knowledge will define the model for a new future for food: a new civilization and a new humanism that we have to build slowly through carefully planned interventions in different sectors.

Sovereigns of Production

To achieve food sovereignty, each people has to be guaranteed the right to produce healthy food, in abundance and accessible to all. Today the number of starving and undernourished people in the world has exceeded one billion; that is to say, roughly 15 percent of the total population. A system that has reached a production output sufficient to feed all the inhabitants of the planet has to make sure that everyone has access to that output.

In the previous chapters, I examined many of the negative processes that have combined to create the present situation. Here I outline the principles that can ensure food sovereignty. I want to demonstrate how, instead of tampering with the dominant world systems and the market-turned-Moloch, what we have to do is act locally on single aspects of the question, correcting where possible the indirect effects of the nondemocratic, univocal global system. Only by working at the grassroots level through the initiatives of communities and local populations, by way of new methods and the recovery of traditional products, will it be possible to secure accessible, abundant, healthy food production for all. Alas, in many parts of the world, this isn't happening, either because the wealthy West has decided it can make do without farmers, or because "enforced" humanitarian aid is destroying local markets and agriculture (this is increasingly the case in poor countries). Another reason is that monocultures and the large-scale projects of international organizations have gained the upper hand. But, though they supposedly herald development, they actually devastate entire food-production systems that once had the merit of at least ensuring subsistence to many people.

Only by giving food communities the power to choose what to produce and how to produce, distribute, and coproduce it (thereby involving the end recipients of their labor, the consumers, too), can we bring to a halt the huge machine that is eating us and the earth together.

Sovereigns of Sustainability

To achieve food sovereignty, products must respect the ecological integrity of the places in which they are produced. A community has this integrity at heart, precisely because it represents its prime means of survival. The food community does not want to jeopardize its natural resources, but rather to make them yield as well as possible. In other situations, where the goal is to constantly boost production through the use of chemical fertilizers and pesticides, whole areas are raped and pillaged, and huge concentrations of monocultures and livestock farms installed. The outcome is that the inhabitants move elsewhere, and soil fertility and biodiversity are compromised for decades to come.

Never in the history of the earth have we witnessed such devastation of the soil and of aquifers. Resources that ought to be public, such as water sources and aqueducts, are being privatized, and the seas are being exploited and polluted.[2] As a result, entire edible species are being put at risk.

Instead of upsetting local economies by imposing on them the concept of development, international organizations ought to focus more of their attention on communities and their existing modes of production. Instead of teaching them how to increase

---

2. The rate of desertification caused by climate change is startling. If we calculate that 70 percent of the water used by humans goes to agriculture (industry accounting for 22 percent and domestic use for 8 percent), and in view of the present demand for an increase in the production of food, it is not difficult to see what the next global challenge will be. Without wanting to, the Bolivian town of Cochabamba has become a symbol of the world water problem. Following the privatization of the local aqueduct, in April 2000, its inhabitants, none of them well-off, saw their water bills go up by 300 percent overnight. All of a sudden, they were spending 25 percent of their income on water. The riot that followed made history.

output to keep up with the needs and caprices of the global "free market," they ought to ensure that communities are effectively working in the interests of their own food sovereignty, flushing out any "wise guys" who happen to be adopting unsustainable practices at the local level. It's a role that would lend undoubted authority and prestige to these organizations. And it's a role they could play without much difficulty, seeing that they have sizable economic resources at their disposal and would certainly have the cooperation of the communities themselves, which have every interest in preventing their lands from being threatened. Not that a respect for the sustainability of processes and the integrity of ecosystems can be taken for granted at the local level. Unfortunately, the temptation to get rich quick often prevails even in small local areas, where it is easy to meet characters out to make money unscrupulously. If communitarian self-regulation failed to suffice, it would be necessary to set up a controlling body, a sort of food-dedicated UN organization, ready to intervene wherever food sovereignty is under threat.

Often farmers themselves are blamed for the advent of industrial agriculture, as if they had mindlessly endorsed it. An oft-cited example is that of the farmer who, following expert advice, spread fertilizers on his land and sprayed pesticides on his crops for the first time. The results were positive, so the second year he increased the amounts, and ended up ruining the quality of his soil. His reasoning was that, "If a certain amount was good for the soil and increased yields last year, this year I'll triple the amount!" None of the salesmen or agronomists who used to roam the countryside promoting fertilizers, pesticides, and the like ever stopped to tell farmers that it was wrong to use more than the prescribed amounts. Farmers were thus deceived and unknowingly became parties to the crime.

The time has come to stop telling farmers how to do their job. The time has come for them to say no to the system and start again from where their fathers and mothers left off.

Subsistence

Primordial agriculture was subsistence agriculture; trade came later. Today, in many of the places described as "underdeveloped," a minimal form of subsistence agriculture still yields food almost exclusively for self-consumption. This is why, to supporters of the market and consumerism, claiming the right to produce food for oneself may sound like a return to the past and the wretched life of the world's poorest peoples; they identify subsistence as a province for the "underdeveloped." Yet growing food in a courtyard or rooftop garden is subsistence too, and it's also true that many farmers who produce food for local or farmers' markets use it to meet their families' needs first before they sell the surplus to others. And who says that swapping the ability to produce diversified foods in exchange for a monoculture or an intensive farm isn't a form of subsistence, too? After all, people who do this might still hang on to a kitchen garden. Who knows?

Speaking of gardens, they are one of the best ways of developing a minimum form of subsistence at all latitudes and in all contexts. With the support of Slow Food and its Foundation for Biodiversity, Terra Madre has given life to garden projects all around the world.

In the Ivory Coast, for example, as part of the Consommons

Ivoirien, Equilibré et Sain dans nos Cantines Scolaires scheme, the Slow Food Foundation has worked with the Chigata Slow Food chapter to promote a consumer-education project in the northern village of N'ganon, fifty miles from Korhogo. Devised mainly for schoolchildren, the project has involved all the villagers and has had positive outcomes for the local micro-economy. It now guarantees the village school two meals a day, featuring traditional Ivorian dishes made from fresh local ingredients.

The story began at a lively meeting in April 2008, when members of the Slow Food convivium presented the project to the inhabitants of N'ganon. The head of the village subsequently agreed to donate a seven-hectare plot of land, which two hundred women pledged to work to provide the school with the ingredients for its meals. Three months later, the seven hectares had been tilled, plowed, and sown with the seeds of the most suitable crops, specially selected by agricultural technicians and engineers.

In the meantime, the women of N'ganon formed a cooperative and, besides supplying the school with food, began to sell their produce at the local market to help fund the project. The one hundred pupils at the school in N'ganon thus eat traditional meals every day, and the economic situation of their families has improved.

This is just one example of the many gardens that have sprung up spontaneously in the recent past, alongside others sponsored by the most diverse institutions. In Italy there are now about two hundred such school gardens, and many Slow Food chapters around the world have promoted this simple form of education-cum-subsistence. From New Zealand to Switzerland, from Germany to the United States, from South America to Africa, the number of school gardens is multiplying all the time.

If this is happening, we have to say a special thanks to people like Alice Waters, the American chef famous for her Chez Panisse restaurant in Berkeley, California. A pioneer of the organic movement in the States, she created school gardens in California to spread the practice round the world and thereby inspired many others to follow her example. Credit must also go to people like young Sam Levin, whom I cited in my speech at the opening ceremony of Terra Madre 2008. Sam planted a garden at his school, involving many of his classmates and supplying food to the cafeteria. Today Sam is much in demand as a speaker all over the United States and abroad, especially in Africa. Hearing him explain why and how he developed his garden, other kids are inspired to follow in his footsteps.

This is where a food system that guarantees food sovereignty has to start; from subsistence, from the right of producers to produce food first and foremost for themselves, from the freedom of anyone to grow plants and raise livestock. All this is possible anywhere, and it doesn't signify a return to the past. It applies to people who live by subsistence alone and to people who supplement subsistence with trade and purchases made with the income from that trade. In other words, whether we are speaking about the South of the world and the most "backward" communities or the wealthy North and farmers who sell most of their food, the argument is equally valid.

Food, by definition, stands for subsistence. If farmers, people directly in contact with the earth and nature, don't practice subsistence agriculture, then the world has little or no chance of returning to reason. Allowing everyone the opportunity to cultivate what they want to eat and what their community wants to eat is not a return to the past. On the contrary, it lays the foundation for a more democratic food system, controllable and capable of yielding a reasonable profit, without infringing on anyone's rights, least of all those of nature: a system in which food is produced first of all to be eaten, then to be sold.

Diversity

As I have said, diversity provides the foundation for Terra Madre, and it is one of the guiding principles that make the network so vital and creative, capable of rallying people at the grassroots level and respecting the identity of others.

Nature itself teaches us that a system with a high rate of biodiversity has more chance of surviving, evolving, and thriving, and is also richer in resources and better equipped to face adversity.

The same applies to people and their cultural diversity, the fruit of a centuries-old adaptation to the area in which they live. Identities are defined precisely as a function of differences: without exchange and comparison, they would be weak, useful only as exhibits in an ethnography museum. For structural reasons, consumer society tends to level out differences. It is its own imperative to survive that requires that consumers have the same needs (or the opportunity to be persuaded that they have the same needs), produced transnationally, in a standardized, one-size-fits-all way. In this kind of framework, in which homogenization reigns, it is impossible to claim the right to food sovereignty based on local ecology and culture. No respect for diversity equals no sovereignty.

Synergy

"Synergy" is a word we often misuse. In marketing jargon it has come to mean generically "working together," a sort of fusion of energies between two parties.

The communities that claim food sovereignty practice synergy as traditional and indigenous communities practice it: synergy among members of the community, between the community and the surrounding environment, between the community and other communities, near or far. In short, the energy exchanged in all these directions is synergy, a hard-to-rationalize, complex system that works and spreads thanks to, above all, affective intelligence. It's not necessary to fully comprehend all this, since it's more than anything else a matter of attitude and respect.

Synergy exists among the members of a community and among the various communities because they share a common store of values. Here freewill giving isn't a heresy as it is in the market economy, where everything has a price. The members of a community and the various communities practice forms of barter, labor exchange, self-help, and fair trade. When these values were propagated in peasant societies, the sacredness of food and labor were never called into question, so it was never a problem if and when strictly personal interests were passed over.

There is synergy with the environment and natural resources because these are not exploited more than is necessary, and their cycles are respected. It is above all through practice that we discover how to make what we have yield as fully as possible. Food too is energy; it represents the sun's energy that, through plants, is turned into nutrition for us. We should exploit this energy conscientiously, with parsimony and judgment, to restore it to the earth in an interplay of the metabolisms that combine to form the "breath of life." To consume more energy in growing our food than the food itself provides in the form of calories is a contradiction in terms that is leading the earth to disaster.

Synergy is the harmonization of the processes described above. And it is eminently achievable at the local level, where it is easier to consider the needs of all those involved in the natural cycle of such a complex cultural and ecological system.

Recycling and Reuse

Food-community systems are oriented toward saving. Waste is antithetical to all their processes insofar as it fails to respect the sacredness of food and everything needed to procure and cook it. Food sovereignty cannot be guaranteed without respect for the principles of saving, reuse, and recycling. Food communities have always shown this respect; it is an integral part of their existence. As I said above, this is one of the main stimuli for a new industrial revolution, which, by necessity, will entail a major process of deindustrialization.

This too may sound like a contradiction in terms, but it isn't. It simply means putting things back into their right place and expunging basic contradictions. Deindustrialization means, above all, reducing scale and centralization and granting more scope to complexity, all of which can be managed by the community of which it is an expression. Local food communities know how to use waste from one process to set another into motion.

In agriculture, the feces of livestock should be used as manure and not compacted into huge masses, full of antibiotics and other harmful substances, that no one knows how to dispose of. Grazing should serve to clean environments and protect them from fire, hence as a means of managing biodiversity. Land grazed by livestock can be efficiently harmonized with vegetable growing. Some animal and vegetable waste may be converted into energy. The technologies I speak of here are new ones: let no one accuse local food communities of looking to the past. By at once slaughtering livestock locally and educating buyers of meat, it is possible to reduce waste and ensure that every bit of the animal is used, not just the prized parts. The gastronomy of any given community is a source of knowledge, of masterly ways of cooking every part of an animal and of using leftovers to make new dishes or, in some cases, as a source of feed for other animals.

Not that the art of recycling should be confined to rural communities. True, it extends beyond the simple, albeit fundamental, practice of separating waste to return it to the productive cycle, but the fact remains that anyone can do it. The pressing need to recycle urban waste is a side effect of the distortions of the consumer system, but this does not detract from the fact that the practice is, in itself, a virtuous one.

To achieve food sovereignty, it is imperative to embrace the logic of zero-waste and zero-disposability, which places the onus on recycling and reuse. In this way it is possible to exit the system, be kind to the earth, and ignite creativity. For, throughout human history, creativity has often worked incredible wonders.

Decentralization

One of the consequences of the industrialization of agriculture and the global food system is an overconcentration and overcentralization of the parts of the supply system.

In the North of the world, countries that thought they could produce food without farmers have opted for concentration, creating large farms that gradually absorb the smaller ones, whose owners, out of either necessity or choice, subsequently abandon the countryside. As table 5-1 shows, the case of the United States over the last century is exemplary.[3]

The same thing has happened in the global South, where the more or less explicit colonization and programmed "development" interventions of international bodies, such as the World Bank or the promoters of the so-called Green Revolution, and the forced introduction of new technology by the agribusiness multinationals have caused a greater concentration of production. The result has been that many small-scale subsistence farmers have abandoned the land or have become employees of large-scale landowners.

Standardizing techniques, monocultures, the need to produce uniform raw materials for export—in the long run, these and other characteristics typical of industrial agriculture have all, irrespective of local geographical context, rowed in the same direction: that of land ownership concentrated in the hands of a few people and mass production.

The same concentration process has taken place on livestock farms. It's worth adding also that monocultures are invariably

---

3. See "Agriculture in the Classroom," http://www.agclassroom.org/.

TABLE 5-1: United States Farm Statistics

Year	Population & Farmers (% of Labor Force)	Farm Data
1900	Total Population: 75,994,226 Farmers: 38% of Labor Force	Number of Farmers (estimate): 29,414,000 Number of Farms: 5,740,000 Average Size of Farms (in acres): 147
1930	Total Population: 122,775,046 Farmers: 21% of Labor Force	Number of Farmers (estimate): 30,455,350 Number of Farms: 6,295,000 Average Size of Farms (in acres): 157
1950	Total Population: 151,132,000 Farmers: 12.2% of Labor Force	Number of Farmers (estimate): 25,058,000 Number of Farms: 5,388,000 Average Size of Farms (in acres): 216
1970	Total Population: 204,335,000 Farmers: 4.6% of Labor Force	Number of Farmers (estimate): 9,712,000 Number of Farms: 2,780,000 Average Size of Farms (in acres): 390
1990	Total Population: 261,423,000 Farmers: 2.6% of Labor Force	Number of Farmers (estimate): 2,987,552 Number of Farms: 2,143,150 Average Size of Farms (in acres): 461

concentrated on the production of animal feed, not directly on food for human consumption. In some countries, the concentration in intensive livestock farms has reached such a peak that living conditions are unhealthy for animals, never mind humans.

The same applies to abattoirs (in the United States, for example, 80 percent of all the country's beef is processed in just thirteen meat-packing facilities) and all the other production and supply chains, which are increasingly forced to comply with the rigid standards of large-scale distribution. The distribution sector itself has undergone a process of concentration and centralization, with large companies gradually absorbing smaller ones. All this is typical of industrial processes, but here it's food I'm talking about. Whether we buy food products at a supermarket chain store or eat at a fast-food restaurant—whose franchises are identical the world over—not even consumption manages to remain immune from the centralization of "food power" in the hands of a few.

To justify an economic model that, in no more than half a century, has turned the world food system on its head, it has been argued that it increases the productivity and, above all, the efficiency of the system; that it allows us to feed the world and control the food that ends up on our tables. But, as we all know, this simply hasn't happened—anything but; the situation has gotten increasingly worse. Only a more decentralized system could make sure that "power" over food returns to local populations and communities with—as many decentralized systems demonstrate—more efficient industrial-production systems.

The Manifesto on the Future of Food cites a 2006 study of more than 200 sustainable agriculture projects in 52 countries, comprising 30 million hectares of land and 9 million rural families. The study, sponsored by a group of universities, shows that sustainable practices—hence relative decentralization—"may lead to substantial increases" in production. Some basic food producers increased their earnings by 150 percent thanks to their sustainable methods. With costs much lower than those of conventional production, organic farmers often earned higher profits, even in the rare cases in which their yields were slightly lower. In general, the yields of organic agriculture proved higher on an "acreage unit" basis. Industrial technicians wrongly take yield per unit of labor as a parameter of efficiency, but in industrial processes most human labor is replaced by machinery and chemical substances, which make a system appear efficient even when it isn't. Industrial output statistics are further distorted by the fact that it is impossible to define the costs of damage to the land, soil, and health of citizens.[4]

Most surveys have demonstrated that small, high-biodiversity farms are just as productive as industrial ones—to the benefit of the system itself. I get angry when I hear people say that Italy is not competitive because it lacks concentrated farms capable of responding to the needs of the market. Here in Italy, in fact, we see before us the positive effects of the decentralization of production and the proliferation of small and medium-sized enterprises. Our many food artisans and the growing and breeding techniques of our farmers provide us products that are better and healthier than industrial ones. We have to say thanks for this to the diversity that comes from the rootedness of our farms in their local contexts,

---

4. See "Manifesto on the Future of Food."

from decentralized agriculture independent of the power centers of seed and feed corporations and large-scale distribution. The most glaring example of the virtues of decentralization is precisely the one that Italian supporters of concentration bandy about as a success model: that of wine.

In 1986, in the wake of the methanol scandal, the Italian wine industry looked as if it was on its last legs.[5] But then, in just over ten years, it enjoyed a renaissance that now allows it to compete in international markets with the finest French wines. What happened? Simply put, many small wineries took advantage of their uniqueness and the distinctive features of their *terroirs* to produce excellent vintages. In next to no time, all the wine producers on the peninsula who followed these strategies increased their outputs, improved quality, conquered sizable market shares, and earned a great deal of praise. Italy was lucky because its various climatic and geomorphological conditions and its vast array of native grape varieties gave it the wherewithal to bounce back without standardizing production in any way. Today the wine world views the standardization of taste as a disaster, and producers are fully aware that it is only by taking full advantage of diversity that they get the best out of their vines and achieve success in the market.

In both the North and South of the world, without decentralized agriculture and without making the most of diversity, with small and medium-sized farms and companies exploiting the characteristics of their own local areas to the full, there can be no food sovereignty—and, one might add, no future.

---

5. In 1986 a few wineries were caught producing wine cut with blends of liquids and methyl alcohol, causing twenty-two certified deaths and doing untold damage to the image of Italian wine around the world.

Biodiversity and Identity

Biodiversity and identity are closely connected. Biodiversity is the evolutionary guarantee that, by way of the principle of adaptation, allows communities to take advantage of natural resources. Biodiversity and the characteristics of a specific locale allow agriculture and agricultural techniques, the ways and means of harvesting and consuming food, along with cooking methods and food rituals, to evolve in unique ways. It is in this way that the identities of peoples and their cultures are formed. We have also seen that identity is defined by another form of diversity, that between the habits and ways of living of peoples. There can be no identity without exchange, no identity without difference.

Identity is always defined by difference. We define ourselves in relation to others; we can't do it on our own. For this reason, we should never feel intimidated by people who are different from us, by foreigners or strangers. Without their diversity, our identity would be threatened.

The same happens in nature. And it is nature that teaches us the value of difference and the wealth of variety. The analogy with biodiversity is evident: without diversity of species, varieties, and breeds, without crossbreeding and selection, nature would be in danger, incapable of dealing with problems such as disease or sudden changes in environmental conditions.

Let us look at a few examples.

In 1970, modern American agriculture had to address a very serious problem when, in the southern states, large plantings of hybrid commercial corn were ravaged by a plant disease known as southern leaf corn blight (*Helmintosporium maydis*). A wild variety of corn in Africa was found to be resistant to the disease, and, thanks to this genetic inheritance, poor Africa de facto saved the sophisticated, ultramodern industrial agriculture of the United States.[6]

Another example is that of the meritorious Vavilov Institute in St. Petersburg, Russia, the world's third-largest germplasm bank, which has been collecting vegetable varieties all over Russia for more than two centuries. It became famous during World War II when, with the city besieged by the Germans, some of its researchers preferred to starve to death rather than eat the institute's precious stocks of grain, corn, and potatoes. More recently, the Vavilov Institute helped solve serious hunger problems in Ethiopia in 1985 and in Georgia in 2000 by supplying vegetable varieties that had been lost in those countries.

Just as defending biodiversity is important to allow natural systems to survive, so our "human system" has to find resources in diversity: biological, intellectual, cultural, social, and economic. If it loses diversity, a system as complex as nature grows so weak that it dies.

Our fate would not be all that different if we all expected to share the same habits, the same ways of creating, communicating, studying, cultivating, eating, and relating to each other. It is no coincidence that we are an integral part of nature with the same right to be in the world as all other living beings. But we forget too often that this is our condition; the truth is that we function just like any other living being on Earth. Blind belief that our artificial models are the solution to any problem can be risky. Occasionally

---

6. Jean-Anthelme Brillat-Savarin, *The Physiology of Taste*, trans. Fayette Robinson, (Adelaide, Australia: University of Adelaide, 2007).

it's a good idea to turn our attention to something we still don't fully understand, something that eludes us, but that works better than anything we have ever invented. I refer of course to Mother Nature. It is she who teaches us that diversity is the greatest vital force that exists, that it has to be preserved.

The secret is to maintain biodiversity and local identities and let them grow as part of the diversity of the world. Each one of us represents a piece of that diversity, which makes sense only as a function of all the others, none excluded. Without biodiversity and local identities, food sovereignty is unfeasible.

Voluntary and Free Trade

When the world's commodity markets fix the prices of foodstuffs, farmers are no longer free to produce and trade. Large-scale distribution throughout the Western world determines not only the quality of food it is prepared to sell, but also prices, which are invariably unfair for farmers.

Food sovereignty empowers the seller to dictate his or her conditions and come to an agreement over price, without impositions from third parties. For this reason, trade should be encouraged inside and among communities, and it should be as direct as possible, with as few middlemen involved as possible.

Commercial initiatives must not be imposed from above, but should be fair and sustainable, mutually advantageous for the food producers and coproducers of communities as a result of free agreements in compliance with their own models and criteria, from barter (and even gratuitousness, hence donation) to bank transactions. Every commercial opportunity must be evaluated only by the interested parties and solely on the basis of its specific merits. No international body can compel a country or community to permit commercial investment or exchange across its borders that might be damaging to local interests.

Trade or commerce performed by middlemen is not an outright evil, but it is hard to find a trader, outside the world of fair trade, who does not look exclusively after his own interests. According to Jean-Anthelme Brillat-Savarin, commerce "purchases at as low a rate as possible what it consumes, and displays to the greatest advantage what it offers for sale."[7] In this sense, it appears to be normal and even fair for the trader not to work with humanitarian or charitable aims in mind. But Brillat-Savarin also said that commerce is part of the economy; in my opinion, this implies that, within a new gastronomy, commerce too should not be devoid of ethical forms of behavior, or at least should seek to be as sustainable as possible in the interest of the community. How nice it would be to set up a "World Company of Honest Traders," who would relinquish a small amount of their profit and donate it to communitarian projects at the service of a world in which food sovereignty would be achievable by all, and in which the local economy would begin to work precisely thanks to their generosity. And, of course, a local trader who feels part of a community is much more likely to be fond of the place he lives in, of its social fabric and shared identity. International traders are not concerned with such aspects and do not live in the places in which they carry out their business. As I said when I spoke of sustainability, what is

---

7. J. Esquinas Alcázar, "Protecting Crop Genetic Diversity for Food Security: Political, Ethical and Technical Challenges," in *Nature*, no. 6 (December 2005).

needed is an international authority to combat trade that is unsustainable, unfair, and undermines the food sovereignty of peoples.

This is why it is so appalling that institutions such as the World Trade Organization (WTO) are among those most responsible for multilateral agreements that help the wealthier nations more than the poorer ones. Such practices should be declared illegal and deemed a violation of human rights.

Seed Patents, Monopolies, and Privatization

Food sovereignty is also the freedom to use the natural resources available at the local level and to implement the techniques and technologies that best adapt to the conditions of the local context.

To allow researchers and the funders of their research to patent forms of life—and hence establish monopolies—is irreconcilable with the principle of food sovereignty. National and international laws that permit these practices are a violation of the dignity and sacredness of all forms of life, of biodiversity, and of the legitimate heritage of the world's indigenous populations and farmers. These principles are valid for all living beings.

Patenting life forms is a negative legacy of industrialism and mechanism, the conviction that nature is at our disposal, a machine like any other. If I make a discovery, this does not mean that the piece of nature I discover is mine; I have only discovered it and figured out how it works, but I haven't invented it. Ethical motives apart, this logic is entirely incompatible with an agriculture that is free and clean, that has to belong to communities without interference from third parties. Patents on life forms deprive indigenous communities of the possibility of using the active principles of medicinal plants (there are many legal controversies over the matter, especially in Latin America), but above all they make the privatization of seeds oligopolistic and spread it globally.

In the last few decades, seeds have been at the center of the greatest robbery the agri-food industry has perpetrated against humanity. Capable of multiplying and reproducing, a means of production and a product, seeds had hitherto withstood the laws of the capitalist market. Research into and development of techniques to improve seeds belonged to communities that, thanks to centuries of exchange, sharing, and practice in the field, had selected the best, most adaptable varieties. Then, as scale increased, more "scientific" approaches were developed and research activity became a prerogative of the public good and a government responsibility. But since the 1920s, the investment of large amounts of private capital, especially for the hybridization of corn, led to the application of patents to the resulting discoveries in order to allow them to be exploited commercially. With the development of genetic engineering in the 1980s, the practice of patenting seeds became more widespread and spawned technological aberrations. One such was the "Terminator" seed, designed not to reproduce, but to yield a single crop, the idea being to prevent farmers from saving and regenerating seeds from one year to the next.

Genetically modified organisms, GMOs, are the product of this commercial system. They are patented, hence are registered as the property of whoever has developed them. This characteristic obviously renders them incompatible with the principles of food sovereignty.

The seed has always been a shared resource, a common good. To transform it into a commodity is to radically change agriculture and the nature of the seed itself. In this way, farmers are robbed of a natural means of subsistence and their livelihood, and technology thus becomes a direct cause of poverty and underdevelopment. To achieve food sovereignty in the world, it is necessary to free seeds from patents and involve farmers in research, which should receive public funding and should be conducted in the fields of the food communities.

GMOs are only the latest and most underhanded of the technologies invented by the private sector to control agriculture. Since certain phenomena can only be measured in the long term, we are not yet in a position to evaluate the real effect of GMOs on human health, on diversity, and on the balance of nature. But insofar as they are designed for monoculture, to own the rights to the seeds that give us what we eat, and to support and add lifeblood to the agro-industry system, this type of technology deserves to be rejected.

Not that GMOs are the only technology developed to further conquer agri-food markets and concentrate them in the hands of a few. Many strategies and innovations—from DDT to synthetic fertilizers—have moved in this direction, proving in many cases to be totally unsustainable and even harmful to human beings and ecosystems.

To secure food sovereignty for all, it is thus necessary to apply a precautionary principle. No new food production technology should be authorized unless its conformity with local safety laws, potential nutritional properties, wholesomeness, and sustainability can be proven.

Existential Sovereignty

The principle of food sovereignty should not be mistaken for a mere political demand—though it is also that—nor for a proposal for a world food policy that would require overview by the appropriate international regulators (though some even say that an ad hoc food authority ought to be set up). Still less is it a utopian protest on behalf of indigenous communities, or other communities deemed "backward," whose freedom to grow and eat what they want in complete autonomy is in jeopardy.

Today food sovereignty concerns all the world's populations. When they go to buy food, every inhabitant of the planet should be able to exercise his or her right to sovereignty, to have the opportunity to choose where to shop, whom to buy from, and to have access to foodstuffs that correspond to all the criteria outlined above. The rules that establish how a food ends up on our tables have to respect these criteria.

Nonetheless, we are not truly free just because we can afford to buy sustainable food that respects our identity and that of the community in which we live, that gratifies us and whoever has worked to produce it. Food sovereignty has to assume a broader relevance. Since food embraces different areas of our existence, it is necessary for the sovereignty concept to play the central role that so far has been denied it, and thus enable us to determine who we are and what we wish to do to live decently and happily on the earth.

Sovereignty of Knowledge Systems

A people's identity is largely founded on the ways its individual members choose to feed themselves. We have seen how the food-community system is closely connected with biodiversity and the characteristics of the place to which the community has adapted (and this applies to any community, not only agricultural or rural ones). The system is so complex that even if only one of the component factors is missing, it may be damaged for good. This doesn't mean that a system cannot change but, as in nature, the arrival of external elements may undermine its existence totally, thus bringing the evolution process to a halt. I am not arguing that systems are unmodifiable, but rather that they can evolve in a sustainable manner.

What ensures the functioning of the integrated system formed by nature, agricultural production, ways of preparing and eating food, social life, and landscape is the corpus of knowledge a community has assembled over the period in which it has inhabited its territory. This corpus includes ancient and ancestral knowledge, but also modern and even postmodern knowledge, such as alternative farmer-to-consumer systems for the distribution of seasonal food over the Internet. Today the delicate balance of this system is in jeopardy. Many communities lack the tools to defend themselves from the pressures of homogenization, waste, the phony convenience of disposability, and irresistible, but never to be fulfilled, needs.

Preserving this precious store of knowledge is a right, but also an obligation—for food communities, for indigenous populations, for all countries—because it is this knowledge that helps to save biological diversity, our heritage of local food skills, and the conservation of common goods to which members of any community must have access without outside interference.

One shared heritage is self-critical awareness of colonization, which subjugated entire populations in the name of a limited, Eurocentric cultural vision, destroying their traditions and plundering their principal exportable resources. In many parts of the world today, industrial agriculture is doing the same thing—no more, no less—though arguably not always with the same violence. The fact that it uses subtler, more silent means does not make it any more acceptable.

The protection of the knowledge of communities, such as those of Terra Madre, is the key to maintaining sustainable agriculture where it already exists, and restoring it where it is no more. No global trade agreement or ownership right can expect local communities to bend to any obligation. No global trade regulation or multinational should be allowed to violate the right of farmers and local communities to possess and pass on their own seeds, their own techniques and technologies, their own rhythms, and their own collective innovation and knowledge. Any form of "biopiracy"—that is to say, the theft of local knowledge and genetic diversity for commercial purposes—should be firmly challenged.

It is only by preserving local knowledge, be it ancient or modern, that communities can have sovereignty over their own existence, and establish research and development goals using their own models with total autonomy.

Economic Sovereignty

Another type of sovereignty that results from food sovereignty is economic sovereignty: the possibility to decide autonomously—from communitarian to national level—what kind of economic activity to practice and what kind of trade and techniques to use for exchange, sales, and distribution.

The financial crisis that hit the world toward the end of 2008, with dire consequences in every sector, showed the true face of a blind, ruthless, speculative approach to finance. It ultimately made money impalpable, increasingly characterized by speed—so much so that it became as slippery as quicksilver even for international money marketeers.

In his book *Slow Money*, an account of his sustainable investment project of the same name, Woody Tasch expressed effectively and cogently the need to restore concreteness to money. "We have to bring money back to earth," he said.[8] He was referring explicitly to the earth, hence to the sustainability of investment—agricultural investment in particular—but also, indirectly, to communities. In other words, investment, profit margins, and resulting interest must somehow benefit communities and remain among the members of those communities.

It's very important to make sure that communities draw an effective benefit from the financial operations that affect them. Above all, peoples must have sovereignty over the administration of the wealth they generate in harmony with nature, without anyone intervening from the outside to take this wealth away

---

8. Woody Tasch, *Slow Money* (White River Junction, VT: Chelsea Green Publishing, 2009).

or erase it, "proposing" or brutally imposing other economic models.

Communitarian economies must not be passed off as marginal or "niche" just because they aren't part of the big picture of international globalization. Unlike the globalized version, these economies ensure that wealth is shared and value added within the local area. Just because they are not functional in the terms of the free-market economy and are incapable of generating profits for those who would like to speculate on them doesn't mean to say that they aren't effective. On the contrary, precisely because they are extraneous to the logic of "rogue finance," they possess the potential to represent a real alternative to the disaster that we see before us.

New Rights and Participatory Democracy

To complete my outline of communitarian and individual rights, I now move to animal rights—for which legal precedents already exist, albeit not entirely effective with regard to animal well-being on farms—and the rights of nature, that is to say the rights of rivers, seas, forests, and biodiversity. These elements, the habitats in which our food is born and grows, must be respected and enjoy the same guarantees as living beings.

The right of rivers to be clean and not overexploited, the right of forests to stay luxuriant, the right of seas not to be polluted (hence respect for their fish stocks) are, of course, rights whose benefits we all enjoy. It is thus possible to include them, albeit indirectly, among human rights. It is no longer possible to consider nature as a separate entity that cannot be defended in the appropriate international contexts under universally recognized legislation.

If the exploitation of these environments and systems is left unregulated, the people who think they can privatize and ravage, exploit and endanger ad infinitum will always come out on top. What with the climate and ecological crises, and in view of the weight that unsustainable food products—the industrially produced stuff, first and foremost—wields nowadays, the rights of nature itself need to be reviewed, broadened, and redefined, and severe sanctions ought to be imposed on anybody who fails to respect them. At stake are the future of the planet, our own future, and the future of generations to come.

Within this framework, we have to show a greater sense of responsibility toward our surroundings, to the places we live in. This is why it is necessary to recover a sense of proportion, this is why we have to become masters of our own existence again, ready to participate actively in all sectors of public life. These are the prerequisites of true participatory democracy.

Today politics is like a foreign body in our society; it is dominated by a caste, and people no longer acknowledge its authority and "representativeness." Just as television does not sell programs but delivers our brains to advertising, and the food industry does not sell food but lets our freedoms and pleasures be eaten by the system, so politics simply uses us for its own purposes. Politics used to be at the service of society, but now it simply brokers votes and swindles the masses. Recovering our existential sovereignty signifies commitment and rules; it means becoming active participants again and not simply disposable voters.

Deindustrializing Food

Food is the key to recovering our lives. It represents a challenge that we have to address globally, at all levels and all latitudes, irrespective of whether a local area is rich or poor, rural or urban, and irrespective too of the quantity and type of food produced there.

People who espouse the line of thinking that has caused us to be eaten by food will continue to accept the present system in the belief that it's the only one possible or the only one worth practicing. All of us, despite ourselves, help keep alive the inhumanity and unsustainability of this system with our acts of production and consumption. But the time has now come for a new humanism, a renaissance that, setting out from the perspective I have outlined, will find simpler solutions than one might imagine to very complex problems.

The aim is not only to adjust the present system, which resembles a train heading toward the edge of a cliff. It is vital rather to invert the relationship that we have and are forced to have with what we eat. This, admittedly, is a very complex operation, but it should be no cause for fear.

In the next chapter we'll examine what steps we have to take. I begin with a description of a very flexible, very open model for restructuring the world of food, a model that promises a new industrial revolution based on deindustrialization.

Deindustrializing the world of food doesn't mean reducing its productive potential or efficiency. It means restoring its natural dimension, in which humans can fulfill themselves and truly enjoy the fruits of the earth. Deindustrializing food means restoring it to those who cultivate it, breed it, or catch it—and those who eat it in a responsible manner. Deindustrialization doesn't mean waging war on the global food system; on the contrary, it means creating alternatives. It doesn't mean yet another system imposed from above, but rather a way of handing over the initiative to food communities and then supporting and assisting them in every way possible. It will be the first step toward building a system of local networks on a communitarian basis, a human system in which food is "good, clean, and fair" again and we become sovereigns of our own lives.

SIX

local economy, natural economy

All we have to do to deindustrialize and revive the global food system is to hand over the initiative to the food communities. It is they who are capable of rebalancing our relationship with the earth; a top-down model would only trigger negative reactions.

The development of a grassroots alternative, taking into account what food communities manage to achieve in their own individual contexts, responds to both economic and ecological criteria. This is because the communities produce and consume sustainably, applying ecological principles to their activities and complying with an original economic model. Theirs is a way of managing "our common home" that has roots in our past, in the history of agriculture and our evolution on Earth, but is also projected toward the future and thus cannot yet be fully grasped.

Local economy as practiced by food communities introduces new parameters to economic philosophy—hard to quantify, but fundamental insofar as they are bound up with the most precious thing in the world, something that can't be bought or sold: the enjoyment of life.

Home Economy

Economy and *ecology* have the same etymological root. *Economy* is derived from *oikos* and *nomos*, *ecology* from *oikos* and *logos*. In Greek, *oikos* means "home," *nomos* "rules of behavior," and *logos* "rational thought." The two words have similar etymologies because rational thought leads to scientific knowledge, which, in turn, guides the rules of behavior in practical life. Rules of behavior plus the home: economy ought to be the set of rules that regulates our home, guided by the rational thinking known as ecology. Economy refers to the way in which we run our home. And if we think of our home in a global sense, then that home is Earth. If economic criteria are not guided by ecological thinking, then the home isn't well run. This is what has been happening ever since we began to think of the economy as a closed system that feeds itself while excluding the nature that surrounds us, namely, our home. It is like living in a house without walls or furniture or windows: a house that isn't a home. How have we let ourselves be guided for centuries by such a glaring oversight? It is necessary to rethink the economy and change the criteria that guide it, basing them around our common home, Earth.

The theories of Nicholas Georgescu-Roegen explain why the present economic system fails to look after the home. His entropy law shows how economic systems overlook nature and yet are ultimately both indebted to it and destroy it. Georgescu-Roegen asks how, insofar as humans are incapable of producing matter or energy, we can produce material things. If we view the economy as a merely physical process, if we view it as a whole, we see that this process is partial, circumscribed by a boundary through which matter and energy are exchanged with the rest of the material universe.

It is a process that neither produces nor consumes matter and energy, but simply confines itself to absorbing them or expelling them, uninterruptedly. That said, economics is more than pure physics. According to heterodox economists, it is a process into which natural resources enter from one end and waste is expelled at the other. Again according to Georgescu-Roegen, if we look at this process from the point of view of energy and according to the laws of thermodynamics, matter-energy enters the economic process at a state of low entropy and comes out at a state of high entropy. The concept of entropy is complicated to explain: put simply, it is the energy that remains "bound"—in other words, an energy we can no longer use. Hence, at the start of the economic process, we have a lot of free "usable" energy, whereas at the end we produce "bound" energy, which we don't know how to use. One might say that every economic enterprise—but also every biological enterprise (since the human being ages and deteriorates in the course of his or her life) is thus bound to have a deficit, to show a loss. That is the way things are, but it should not discourage us from undertaking the things we choose to. Nonetheless, we have to realize that, "The true product of the economic process is an immaterial flux, the enjoyment of life."[1] This turnaround alters our perspective a lot.

Here I have briefly summed up Georgescu-Roegen's complex theory to show how, when it comes to economic processes, we often pick up the wrong end of the stick. We should seek, above

---

1. N. Georgescu-Roegen, *Energy and Economic Myths* (New York: Pergamon, 1976), foreword.

all, to minimize entropy, meaning the production of unusable waste or energy, or at least seek to allow ourselves and the earth to enjoy this "dispersion," turning it into usable energy for natural processes, their functioning, and their regeneration.

To make this possible, we have to create more decentralized, manageable systems. If we recover a holistic orientation to production and natural resources, to conservation and utilization of resources, to the social and cultural aspects of the question, we shall also rediscover the true quality of life.

With this prospect in mind, it is possible to activate "hidden connections" and make them work together for the common and personal good and to achieve the balance required to keep the system standing.

Every community should be given responsibility for looking after its habitat, be it urban or rural, and the first thing it should pay attention to is food, the most direct link with the "home." This does not mean shutting ourselves off from the world or complete self-sufficiency, but rather awareness, responsibility, and participation. In the globalized world, a whole host of possibilities open up. The first is the interconnection of the different systems of local economies. Networks can be created to enable the movement of material and immaterial goods: from products as such—in a sustainable manner and wherever the need is real—to information, to the sharing of knowledge and useful technologies, not to mention the convivial aspects of the immaterial but fundamental "enjoyment of life."

The Advantages of the Local Economy

Production

Local production is based on the cultivation of local vegetable varieties and/or the raising of animal breeds that are as far as possible native, that have adapted in the course of time to climatic and geomorphological conditions and the characteristics of the land, continuing to evolve in harmony with the area and with the populations (human and otherwise) that live there. Precisely thanks to this adaptation, such species are more resistant than others and do not need to be treated with antibiotics or hormones and given feed grown with chemical fertilizers and pesticides.

Local production also helps maintain agricultural tradition and knowledge. Local foods grow better if they are cultivated according to tradition; though tradition does not exclude innovation, rather, it evolves through a combination of nature and the manual dexterity of humans, without upsetting a type of agriculture that has been practiced for centuries.

Culinary traditions also stand to benefit because local food is obviously best processed locally. Yet another advantage is that local production helps small- and medium-scale growers to live decently with an income commensurate with their labor, and without being crushed by the mechanisms of the large-scale system. These growers either work for a local market and its outlets, or sell traditional products with high added value outside the local context.

One example of a Terra Madre food community/Slow Food presidium is that of the Dogon people of Mali, who inhabit the Bandiagarà escarpment, which runs from north to south, from Timbuktu to Mopti. Here they live in caves hewn into the red sandstone rock and in mud huts. When the first anthropologists began to study their culture, the Dogon, a people of blacksmiths, farmers, and healers, amazed the world with their extraordinary knowledge of astronomy. Without the aid of technology, they knew of the existence of the inner ring around Saturn, the four satellites of Jupiter, and the invisible star system of Sirius. Now the Dogon are famous throughout Mali for their skill as onion growers. Their many traditional dishes include millet and bean beignets, *tò* (millet polenta), shallot and wood-sorrel patties, powdered onion and baobab, *acasà* (peanut and sugar patties), millet couscous, and millet beer. The only food product the Dogon sell commercially in any quantity is the shallot (fresh or dried). The most interesting feature of their agriculture is the variety of crops they grow in their traditional gardens, in which areas are allotted to fruit trees (mango, orange, banana, and karite, or shea), cereals (rice, corn, millet, and fonio millet), peanuts, vegetables, and legumes. Together with the knowledge of their women, who use the flowers, fruits, and leaves of every plant (cultivated or wild, such as the baobab) to make a special condiment, so much biodiversity in such confined spaces (sometimes gardens are as small as a hectare in area) is without doubt the Dogon people's most precious resource.

In Africa in particular, local agricultural production is seriously menaced by the global market, which is weighted in favor of the world's North, where the wealthier countries heftily subsidize their own agricultural produce. Partly due to these subsidies, imported produce in Africa often costs less than locally grown. Thus, a little at a time, poor countries increasingly depend on food imports.

Senegal, for example, imports 95 percent of the rice it consumes from Thailand, whereas traditional rice production in the north of the country has no market outlet. What's more, rice consumption has de facto replaced that of native cereals—of fonio millet in particular—thus revolutionizing the Senegalese diet and causing a progressive reduction in local species and varieties.

With food grown locally, at worst people always have access to a minimum level of subsistence and the possibility of exercising their food sovereignty. And the earth will always provide sufficient resources to solve any type of crisis and problem that may emerge.

Distribution

The local economy allows us to sidestep the middlemen in the supply chain. It is easy to develop local and regional agriculture in cities and invent forms of direct sale in the countryside. Exemplary in this sense are farmers' markets and consumer groups in all their shapes and forms: from those whose members buy directly from growers to the ones that order food on the Internet, with producers delivering goods at a collection point or directly to buyers' homes. I have to give credit to the United States, arguably more famous for being the nation that invented fast food, for inventing farmers' markets and community supported agriculture. Or, rather, in the States they reinvented a type of market that used to be common throughout Europe, before farmers and producers were replaced by mere traders. The rediscovery of seasonal, fresh produce, the possibility for farmers to earn better incomes with greater professional satisfaction and a renewed town-country relationship came as a reaction to industrial agriculture and its poor-quality products. The farmers' market phenomenon has since enjoyed a boom in the United States and spread in areas in which it had previously disappeared. Today there more than 4,600 farmers' markets in the United States, even in the major metropolises, where consumers have access to fresh, seasonal produce as though they lived in the countryside.

Inspired by Terra Madre, Slow Food has launched the international Earth Markets project and is developing consumer groups that work both autonomously and in collaboration with specialized associations.

Earth Markets have so far been set up in Tel Aviv in Israel, in Bucharest in Romania, and in Beirut, Saida, and Tripoli in Lebanon. In Italy a pilot project was developed at the Mercatale market at Montevarchi, in Tuscany, and was followed by others in Bologna, San Daniele del Friuli, and San Miniato, near Pisa. These markets are now being held regularly, and similar initiatives are multiplying all over the world, especially where food-community producers already run markets.

The Earth Market project in Lebanon comprises three different areas. In the north a weekly market is held in Tripoli, the second most important city in the country. In the south, where the conflict with Israel has caused the worst damage, a market has been set up in Saida, once a Phoenician merchant town. Beirut, the capital, is the natural venue for the third market.

The Saida market, inaugurated on April 6, 2008, is held every Sunday in the Khan El Franj, a magnificent old building on the town's seaside promenade. Its stalls sell fresh fruit and vegetables, *mouneh* (typical Lebanese preserves), and artisan-made traditional cakes, spirits, honey, olive oil, and natural soaps.

The El Mina market opened in 2007 and was moved to Tripoli on August 26, 2008, with the aim of reaching a wider public. It is held every Thursday, and is attended by about ten producers of preserves, fresh fruits and vegetables, and crafts.

The Beirut market was inaugurated on January 20, 2009, in a side street off Hamra Road, one of the city's main thoroughfares. It is held every Tuesday and allows fifteen producers to sell their foodstuffs directly alongside associations and cooperatives representing small producers who could not otherwise participate in the market.

In Israel, work on converting the Tel Aviv farmers' market into an Earth Market began in the course of 2008; as was the case in Lebanon, it was necessary to adapt protocols to local requirements. The market opened on May 16, 2008, and, when the conversion process was completed, in February 2009, it was officially declared an Earth Market.

Situated in the new port of the Israeli capital and held every Friday, this is the first Israeli producers' market. It hosts about thirty stands and a great variety of products; from olive oil to wine, from sheep's-milk cheeses to beer, as well as fresh fruits and vegetables and Middle Eastern classics such as tahini, the sesame paste used to make hummus, and cakes.

The market is located in the recently redeveloped port area, which is packed with bars and restaurants and is very busy, especially during weekends—Tel Aviv being a young city of great cultural ferment. Customers come from all social strata, and at the market they find high-quality produce (many restaurateurs seek out the very finest in the early hours of the morning) at very reasonable prices by city standards. Producers, too, come from all sorts of backgrounds: in a complicated country like Israel, the fact that there are vendors of different religions from every area is particularly significant. This type of activity reconstructs the relationship between town and country, one of the key aspects of the process of redefinition and reorganization of the global food system.

For all these examples of distribution, the local economy is fundamental. Besides offering the evident dual benefit of fetching better prices for producer and buyer alike, it also enables the two parties (producer and "coproducer") to be close to one another, to meet and know where each other lives. If I know the farmer who produces my food personally and hence know how he works, I am thus able to control the quality and wholesomeness of his produce directly, to ask for information and demand improvements and complain if necessary. Direct acquaintance is the best guarantee for good quality and may also foster important interpersonal relations. Farmers feel less isolated and have a clearer idea of the social utility of their work. Coproducers feel actively involved, feel surer about the food they buy and eat, and are able to educate themselves and their children about agriculture and country life in general.

Consumption/Coproduction

The local economy can transform the act of consumption—impersonal, wasteful, and never truly satisfactory—into an active choice, as a result of which the consumer becomes a coproducer. The physical or virtual proximity of producers and the places where production takes place helps us to feel involved in the process that brings food to the table, promotes the circulation of information, and teaches us to appreciate food that is vastly different from the stuff that comes to us via the channels of the world agrifood industry. In 2008, according to data published by the Italian National Institute of Statistics (ISTAT), the direct sale of agricultural produce increased by 8 percent: the coproducer's sensitivity is growing, and the pursuit of local food is beginning to appear not so "childish," as some would superciliously have it, but rather a matter of common sense.

Outside the local context, it is difficult to become a "coproducer." Information on products and producers is hard to come by because of industry's tendency to keep processing methods and the composition of raw materials secret. This is the case with flavorings and aromas (to be found in our food, but also in designer fragrances), whose formulas are jealously guarded and protected as industrial secrets by the five companies that de facto monopolize their world production.

In the local context, consumption becomes the final act of, and no longer separated from, the production process. Instead of destroying something, in its own way it constructs something: it makes a proper use of energy (food is energy), it fights waste (if I can buy food in the vicinity, I won't run the risk of buying too much), and it educates us about seasonal cycles and the characteristics of native as opposed to commercial food products. In a local economy, consumption becomes responsible. And, if resources are not being utilized as they ought to be, perception of the fact is immediate.

Coproducer and producer together are the guarantors of the integrity of the area in which they live and work, and whose fruits they can enjoy—if they manage that area responsibly.

New Gastronomy: The Science of Happiness

The new gastronomy places food at the center of a renaissance in the concept of well-being and a new humanism. It is based on a new approach to agriculture and, through multidisciplinary science touching on all aspects of human nutrition, cannot be separated from the local economic context of food and everything connected with it, however indirectly.

Only local economies and systems—the one we live in, first and foremost—enable the new gastronome to exercise his or her will and to take part as a coproducer, to study and be informed, to have access, by means of travel, to food/human/land systems that are intelligible and integrated and whose survival is thus guaranteed.

This new gastronomy is a philosophy of life, a way to restore food to a central role and make it the pivot of an alliance with the planet we live on and the nature that inhabits it with us. I once defined gastronomy as "the science of happiness" because it blends the old with the new, scientific experience with belief and sacredness, respect for others with self-love. Its ultimate aim is to allow us to live well in harmony with our surroundings, to enjoy honest pleasure and learn the meaning of the world once more. Our senses are the key to that meaning; we have to use them to understand, enjoy, choose, help our neighbors, and interpret the reality that surrounds us. To stop tasting means to stop knowing. If the delegation of culinary, producing, or agricultural skills to someone else is not a knowledge-based act of faith, then it is a hazard. It means yielding up a part of our lives to strangers who don't always work for our benefit; on the contrary, they likely only have their own interests in mind. If fiduciary relations fail, it is hard to figure out where we stand and what our role is; it is equally hard to maintain a solid, healthy relationship between our own needs and concerns and those of a boundless, dying Earth.

It is at the local level that we can practice the new gastronomy;

it is at the local level that we can give a meaning to our presence on Earth. Without this base, much of what we are and what we do makes no sense. Losing savor, we lose wisdom.

Sustainability

In a local economy, even sustainability stands to gain. Distribution methods that cut out the middlemen and allow food to travel shorter distances also reduce CO_2 emissions and all the pollution caused by transportation.

Food requires fewer chemical inputs—from fertilizers to pesticides to preservatives—so fewer foreign substances are introduced into nature. Low- or non-intensive crops need less water. A livestock farm that isn't of mammoth proportions and is suited to local requirements doesn't need to release into the environment effluents poisoned by antibiotics and ammonia, or any of the other harmful substances characteristic of liquid waste from industrial farms.

The protection of an area's biodiversity can have positive spin-offs, one such being the conservation of the landscape and the aesthetic and functional identity of an area.

This has been the case in the Albtrauf, the northwestern edge of the the Swabian Alb, the low mountain range in the southern part of the Baden-Württemberg region in Germany, in which Champagner Bratbirne perry is made. The area is a treasure trove of rare heirloom fruit varieties: pears, plums, apples, and cherries. The fruit trees and the rolling hills combine to create the greatest Streuobstwiese (or "community orchard," from *streuen*, "to scatter," *Obst*, "fruit," and *Wiesen*, "meadows") in Europe. The perry produced from the Champagner Bratbirne, an ancient variety of pear, is protected by a Slow Food presidium, but the true purpose of the project is to preserve the area's magical landscape.

Another example is to be found in Great Britain, where the Slow Food Three Counties Perry Presidium promotes a traditional English beverage made from the fermented juice of pears grown in Herefordshire, Worcestershire, and Gloucestershire. Trees begin to bear pears suitable for perry-making only after a few decades of growth; indeed, the best may be over a century old. The old pear orchards, with their magnificent tall, twisted trees, are an important feature of the British landscape and form an ecosystem regarded as totally unique by British naturalists.

Biodiversity is essential to the local economy, and this is why we have to make sure that it is protected and utilized in a sustainable manner.

The consumption of seasonal products is not only pleasurable and educational, it is also necessary to make production efficient, in harmony with and to the benefit of natural cycles. The reduction of the invasive technologies typical of the agri-food business also has a beneficial effect on ecosystems and natural processes.

The efficiency of productive processes is measured partly in terms of the negative effects they may have on the land. Where controls are stringent, locals are in a position to report anyone who pollutes the environment or implements unsustainable practices. It is better to control a small area and have it controlled by the people who live there, who have every interest in preserving it, than to banish food production from the community and concentrate it in remote areas, where there are no longer farmers or inhabitants, and where anyone can do whatever he wants to without being observed.

Energy

There are other points in favor of local energy systems, in terms of both consumption and production. As I have said, a local food system tends to be "organic" and depend on solar energy to produce food.

The production and use of fertilizers and pesticides demand a huge consumption of energy, which de facto replaces the solar energy that ought to enter the food-production cycle. Industrial agriculture also tends to be "thirstier" than organic agriculture, and thus uses up more water.

The most interesting feature of local-economy systems is that they promote the decentralization of energy production, entrusting the job to individual companies or small local consortia that can make extensive use of alternative and renewable energies. Installing a waterfall or solar panels can cover the energy requirement of an entire farm, and, in the event of a production surplus, new clean energy can be made available for others to use. And let us not forget biomass conversion, whereby production waste and vegetable or organic material that would otherwise be thrown away is used to produce energy. Small biomass generators can in fact meet a farm's energy requirements without the added costs of waste disposal and outside energy sourcing.

The decentralization of energy production is at least as important as that of food production because it permits a more rational and complete use of local resources, as well as a significant reduction in the waste generated by centralized production and the transport of energy over long distances. If it does not interfere with food production and takes place in advantageous conditions, the local production of biofuels may also make sense. Growing certain biofuel crops on land that is abandoned or unsuitable for food production and in fields under rotation can even help regenerate the soil.

Biofuels are fuels derived from living organisms or the waste thereof, and research is concentrating on two main groups: biodiesel and ethanol. Biodiesel is produced from vegetable oils: soy, canola, palm, coconut, peanut, and sunflower. It can be used both as a substitute for diesel oil and blended to make a fuel that can run existing engines without any need of alteration. It is completely biodegradable, and its yield varies from crop to crop (by way of example, palm oil's yield is excellent, soybean oil's is poor); in general, it has an energy content equal to 90 percent of that of diesel fuel. Europe is the world's main producer, partly by virtue of the fact that the Continent has set itself the objective of covering 5.75 percent of its energy consumption with biofuels by 2010.

Ethanol, also known as bioethanol, is an alcohol produced by fermenting agricultural crops with high levels of carbohydrates and sugars; that is to say, cereals (corn, sorghum, wheat, barley), sugar crops (sugar cane and beets), fruit, wine marc, and potatoes. It may be blended in a 3 to 7 ratio with traditional gasoline and used without any need to modify engines. Ethanol can also be extracted, through enzymes, from the biomass of wood and agricultural by-products, but in this case it is more expensive to produce. Another opportunity that has not yet been fully exploited is the use of plants such as switchgrass (*Panicum virgatum*) or a variety of miscanthus (*Miscanthus giganteum*), a native of China and Japan, which, being inedible themselves, may grow on land unsuitable for food crops. The energy content of ethanol is

67 percent that of gasoline, and world production is concentrated in Brazil (from sugar cane) and in the United States (from corn).

But are biofuels sustainable? To answer the question it is necessary to compare their yield per hectare in fuel with their energy yield. From these data, we see that Brazilian sugar cane yields about 8 units of energy per unit absorbed, whereas American corn yields 1.5 units. So why did the United States in 2004 alone cultivate 32 million tons of corn to produce 3.4 billion gallons of ethanol?

It should be pointed out that yield is not the only yardstick for evaluating the sustainability of biofuels. Those in favor of biodiesel argue that it is sustainable because, during their growth process, plants absorb as much carbon dioxide as is released into the atmosphere when they are used as fuel, and also reduce the emission of fine particulates by 65 percent. The use of ethanol, instead, would ensure an 80 percent reduction. Yet these calculations fail to take into account many other factors; for example, the emissions biofuels generate during their production (1 kilogram, or 2.2 pounds, of fertilizer releases 7 kilos, or 15.4 pounds, of CO_2 into the atmosphere), refinement, and transportation.

Considering these data, and bearing in mind how much ethanol "burns" during production—though its energy yield is only 70 percent that of traditional fuels—some scientists have gone so far as to claim that it only reduces emissions by 13 percent.

To make room for crops for biofuels, whole forests are often cut down. This happens in Brazil, but also in China and Indonesia. Some argue that, to minimize the net flow of CO_2, it is preferable to maintain existing forests and grow new ones on cultivable areas rather than produce biofuels from agricultural crops.

In the final analysis, given its high energy yield and value (which minimize its negative effects), arguably the only reasonably sustainable biofuel production is that of ethanol in Brazil. So are we to turn Brazil into the world's fuel tank? Would it be sustainable to transport these new "gasolines" around the world? And, to produce them, are we to cut down huge sections of the Amazonian rain forest?

The situation is further complicated by a whole series of other, by no means secondary, questions. Owing to the "ethanol rush," the price of U.S. corn has increased by almost 200 percent in the last two years. In the same period, people have rioted in Mexico in protest against the rise in the price of tortillas, which are made from corn. How sustainable is it to take space and resources away from food production to produce fuel? Is it morally acceptable that, with the amount of cereals it takes to fill the tank of an SUV, we can feed a person for a whole year?

Energy yield, deforestation, food security, political interests, agricultural policies, GMOs, industrial lobbies—if we consider all these factors, calculating the sustainability of biofuels on a global level becomes a maze in which it's easy to get lost. As in many other cases, the challenge can be won, if we address it at the local level—the only one at which it is still possible to apply common sense and calculate to a certain degree of precision the sustainability of production processes and consumption. It is only at the local level that, in the event of need, we can give priority to food over fuels and make use of the most economical resources (waste and garbage included).

Reuse, Recycling, Saving

Food communities practice reuse, recycling, and saving. In every family, these are the three conditions indispensable for good home economics. They should also apply to the "economics" of our common home, the earth.

Managing saving, reuse, and recycling is a problem of system control, and it is to be presumed that local populations are able to control their own systems with relative facility and efficiency. But reuse, recycling, and saving are also practical actions; they involve know-how that needs to be taught and learned, and they make use of techniques and technologies—some ancient, others brand-new. In local contexts, this knowledge has been shaped over the centuries with adaptation to the characteristics of the local area. From this point of view, practice (meaning popular wisdom) makes perfect, as the brilliant inventions it has come up with in the fields of energy and food testify. I have spoken of cooking and of recipes that use every part of an animal or plant. The same approach is to be found in the production of domestic energy and craftwork. In the course of time, many artifacts have developed into veritable craft products, symbols of a given area: cow's-horn knife handles and other utensils come to mind. The local economy encourages people to sharpen their wits and make the most out of what they possess; and I say this without any intention of returning to the past, when the driving force behind creativity was, more than anything else, shortage and poverty. It is, however, possible to take examples from this past to understand how to set sound, ethical processes into motion in order to generate income or subsistence without repercussions for the environment—indeed to help save it.

Reuse, unlike recycling and saving (which are nonetheless values of fundamental importance), is arguably the most relevant thing you can do today. It is all the rage in every area of culture:

from music to cinema, from fashion to literature. Thanks to the Internet and its fast, efficient search methods, it is the stylistic leitmotif of the century so far. Unconsciously or on purpose, we reuse concepts and ideas and anything else that is immaterial. I don't see why we shouldn't do the same with material things, which are a source of pleasure, feeling, and happiness. As Don De Lillo has written of things as well thoughts, "The world is full of abandoned meanings."[2] I don't see why we can't recover the meaning of life and set it on a new tack.

Participation

In a local economy, every process in the system involves citizens and charges them with responsibilities. This applies not only to the productive sphere, but also, for example, to the conservation of the landscape, to natural and architectural beauty, as well as to more immaterial things, such as tradition. This responsibility is not imposed, but assumed willingly, the local population being aware that it is managing its area on its own and is an active part thereof—even in the simple act of eating. The local economy restores us to our place, our identity, and our existence. This is why political and economic powerbrokers fear it like the plague. It is no coincidence that Charles De Gaulle asked how anyone could govern a country that has 246 different kinds of cheese! Diversity arouses fear because it can't be controlled from above.

If we have faith in communities, which have to act primarily for their own well-being, we can still hope to govern our own lives again; in such an intentional community each member assumes

---

2. Don De Lillo, *White Noise* (New York: Viking, 2005).

responsibility for the piece of the world in which he or she has had the good fortune to be born or to live.

Memory

In a society that has no qualms about wasting things, knowledge and tradition, the elements that define identity, are also wasted. Local economic systems, on the other hand, preserve memory; they are obliged to because they know that their memory and the memory of their community are indispensable. By preserving memory, they take care of themselves. If I have memories connected to something—a sofa in my house, for example—it's easier for me to take care of it, maybe having it repaired and keeping it, than to get rid of it. Memory means care and attention, and local memory means care and attention for places, seen not only as physical sites, but also as interconnected economic, food, and cultural systems.

It is the community's job to preserve local memory. It is the community that has to study and to protect that memory, but also to prolong it, handing down agricultural practices and social conventions, as well as ways of eating and the convivial rituals tied thereto.

The preservation of local memory is a typical example of how the old and the new can interact in postmodern society. It's easy to record the stories of elders, traditional skills and practices, modes of production, and ways of eating. A micro-history of characters and of their little or great achievements can be found in every community. The struggle for survival and the sometimes ingenious ways of making ends meet represent on their own a past that has shaped the present and the appearance of the places we live in.

To forget this is to lose our identity and the profound meaning of our lives and the places we inhabit.

This is why I decided to set an example and found the Istituto Storico di Bra e dei braidesi (Historical Institute of Bra and Its People). I appointed a young local historian to run the institute, and an active group of supporters and sympathizers has since grown up around it. We manage to preserve something of our local memory by publishing a magazine and organizing public presentations of each issue, which attract a surprisingly large number of people. The magazine delves into the history of the town and its inhabitants, and publishes photos and documents sent in by families. Insofar as we are dealing with micro-history and involving people who know or knew each other, our approach is informal as opposed to academic. We shoot video interviews with old people and the characters who appear in the stories we tell and, little by little, we are gradually putting together a precious archive of the history of our town.

In the Terra Madre program, we advise communities to set up associations and groups to address the question of local memory, to record history so that it isn't forgotten, to recount how their ancestors used to live. Many communities are taking an interest and enquiring about our project in Bra. Some have found that similar institutions to ours already exist in their local areas and have thus already embarked on this kind of initiative. Ours is more than just a trip down memory lane; in a two-hour video in which an old man simply tells his life story, even the minor details can enlighten young people, lighting a creative spark and encouraging them to attribute new meaning to the present, to feel new stimuli, greater self-awareness, and pride at being part of a great collective history.

In the context of a network of local systems, if each community preserved its own memory, it would be possible to create exhaustive data banks of details and stories and techniques (which, despite appearances, are anything but obsolete)—of ways of living capable of restoring happiness. These data banks would be of immeasurable value, effective bulwarks against the spread of homogeneous "globo-thought."

Intergenerationality
In a society that wastes things and appears to live for novelty alone, everything old loses importance and can be sacrificed without regret. "Old" are products at the supermarket past their sell-by date; "old" is a product after an analogous but improved version has been launched; "old" is technology with an exponential degree of obsolescence. "Old" too are the elderly. It's worth recalling the sacred role of elderly people in many traditional and rural societies. How are elderly people treated today? As a nuisance to be hidden out of sight, to be locked up in rest homes, to be put in the charge of a professional caregiver. In social terms, this way of dealing with old age—no longer viewed as the age of wisdom but rather as the age of decay and uselessness—has cancelled communication and contact between the generations almost entirely. Take our gastronomic memory: it was formed above all by our grandmothers, who taught us to eat and savor food, cooked dishes that we'll never forget, and defined our taste forever. I can't imagine where the young generations could find better teachers, better educators.

This is why I think that one of the duties of local economic systems is to take responsibility for intergenerational relations and create opportunities for knowledge transmission between old and young. With the depopulation of the countryside, the average age of the rural population is increasing all the time—but, alas, there's no one left to gather its legacy. Young people cannot return to the countryside or engage in food production without the help of the old people who have hitherto guarded local economies, in which memory and modernity fuse together naturally and appropriately. There are already examples of this in Terra Madre communities and in Slow Food chapters around the world. One such is "Grandmother's Day," introduced by Slow Food Ireland and celebrated on April 25, when kids spend their time in the kitchen with their grandmas. In one of the many competitions that are staged, children present their favorite recipe, the one they like to cook with their grandmothers. In another, an art competition involving six Irish schools, children have to draw or paint a picture of "Cooking with Grandma." Grandmothers themselves also have the opportunity to speak about their childhoods and explain their recipes. In Ireland they are now planning to repeat the experience once a month and are organizing an "International Grandmother's Day."[3]

In Italy, as part of the "Orto in Condotta" school-garden project, grandparents teach schoolchildren how to grow fruit and vegetables and, through courses, videos, and public demonstrations, housewives from Modena teach them how to roll out pasta dough to make *tagliatelle* and tortellini.[4] Their manual skills would otherwise be lost, as would the flavors of their cooking.

---

3. *Slow Food Times*, online newsletter: http://newsletter.slowfood.com/slowfood_time/09/ita.html#item2

4. See http://educazione.slowfood.it/educazione/ita/orto.lasso; and http://rezdore.provincia.modena.it/cantiereaperto.asp.

Yet it is undeniable that we can and must do more to foster intergenerational relations. We have to respect the elderly, who, since they have so much to offer, are integral to the local economy. We are living through very delicate times: the generation that is growing up now is the first to have the umbilical cord with its grandparents cut. When these grandparents die, the young people of today, a generation that is getting ready to govern, risk feeling more lost than ever—and for a variety of reasons.

Holistic Vision

We have seen how the local economy offers numerous advantages with respect to the dominant production system. These advantages transcend the food system as a whole, but are indissolubly linked to it. The local context should in fact be interpreted holistically and systemically, as an entity entrusted to the skilled hands of local communities, which, far from dividing, tend to unite their members and to network to solve any problems. From the point of view of production, they do so through innovative distribution systems, in which the new figure of the coproducer replaces the consumer and practices a new gastronomy, all in a sustainable way with decentralized and efficient forms of energy production. Local-economy systems are founded on food and make it the center of a new humanism and a renaissance in which we take control again of our lives and the places in which we live. What count in this context are hidden connections: music, architecture, languages, and other expressions of culture and identity all become important and significant, an active part in a system that creates as opposed to destroys. This vision is interdisciplinary and overrides barriers; above all, it realizes that the energy that is inevitably dispersed during a creative or productive process has to be harnessed to the enjoyment of life, without undermining the balance of nature and the very existence of the earth.

The Terra Madre communities possess this holistic vision, so they aren't surprised when we ask them to play their traditional music or to explain their dialects and thereby help save the world's linguistic diversity too. The Terra Madre communities are like this because this is the way they have always been; they reflect an old mode of doing and making things that has come back into vogue. Or else they are like this because they have understood that the industrial-consumer road was not the right one to follow, and they have delved into their past and their memory to come up with new forms of local adaptation. The local economy is not a utopia; it already provides a base to work on and work with. It is the most versatile tool at our disposal, if we are to emerge from the crises we face and build a better food system, one that radiates happiness throughout the world.

The Network

The communities are not isolated; local-economy systems are not closed in on themselves. In fact it is absolutely necessary for systems to be open to the outside world; if they are not, everything will collapse and the local-economy idea will only work on paper. The Terra Madre experience has demonstrated that local economies are stronger and more authoritative when they are structured as a network, when they become intercommunicating nodes, when they allow people to connect and travel. An example of what the network can generate locally through exchange, travel, and mutual assistance is the Imraguen Women's Mullet Bottarga Presidium, also a Terra Madre food community, in Mauritania. The presidium came into being in 2006 to promote the *bottarga* (dried mullet roe) produced by these nomad fisherfolk, who follow the movements of the huge shoals of golden gray mullet and ombrine along the coasts of the Banc d'Arguin, an extraordinary natural marine environment abounding in fish, protected by a national park set up in 1976.

With the mullet roe the women prepare traditional *bottarga*. When the presidium first came into being, the work still took place in poor hygienic conditions, mostly on dry land in impromptu drying rooms, and the *bottarga* was being bought by middlemen at ridiculously low prices.

Thanks to the Slow Food Foundation, whose main aim is to source funds for biodiversity-protection projects, the members of the historical Slow Food Orbetello Bottarga Presidium in Orbetello in Tuscany became involved. In conjunction with the food technologist Augusto Cattaneo, they organized training courses to help forty Imraguen women improve their processing techniques. With the support first of the Tuscany Regional Authority, then of the Piedmont Regional Authority, a standard production workshop was rented and fitted out and equipment was bought to vacuum-pack the *bottarga*. Last but not least, the wholesomeness of the product was monitored at the Turin Chamber of Commerce chemistry laboratory.

On more than one occasion, at the Slow Fish event in Genoa and at the Salone del Gusto and Terra Madre events in Turin, the women of the presidium have presented their excellent *bottarga*, properly packaged and labeled.

The Imraguen women's next objective is to buy a plot of land on which to build and outfit a workshop. Here they will organize training courses to involve new producers and the fishing community in general. A longer-term objective is to develop a salt pan at Nouadhibou to ensure an all-local guaranteed production and supply chain. To make all this possible, two food communities, one in Italy and the other in Mauritania, have exchanged visits and worked together: a case of the "strongest" helping the "weakest." The project was funded by institutions (though the sums required were by no means high; in 2008 a total of just over 3,000 euros), while all the Slow Food Foundation for Biodiversity had to do was set the machine into motion and organize the exchanges between communities. From a technical point of view, the project has improved a specific food product while respecting its original characteristics. But it has also done much more than that, providing an opportunity for the Mauritanian women and the Tuscan fisherfolk to communicate, to make friends, and to see new places.

In a network it is vital for information to circulate, for the new fair trade I have spoken about to be practiced, and for people to have the guaranteed right to travel. Without the network, many of the values and advantages of the local context risk being lost for good.

Contact and mutual acquaintance add value to diversity and create identity. Meeting others makes us proud of ourselves; meeting others we give and receive help. A local-economy system may appear somewhat limited, but the experience of Terra Madre teaches us that its many possibilities can be multiplied through contact with other communities in every part of the world. At times, indeed, real contact is superfluous; all that is necessary is an awareness of each other's existence.

The network not only enables communities to have access to resources that otherwise might be lacking, it also acts as a multiplier of every appropriate process that is implemented at the local level. After all, on reflection, food itself is a network, the result of many different processes implemented by many different people, of their knowledge, and of the place they live in. It is thus no coincidence that local economies start to network over matters of food. They are the reflection of food and encapsulate all its value, all its meaning.

Our network of strong, self-aware local economies—the network of Terra Madre and all those who are, and could be, part of it—is the greatest food organization in the world today. It excludes no one, not even those actors we might now describe as "unprincipled" by our lights; for, if we embrace complexity, it's better for us not to set ourselves limits. Terra Madre is the true food multinational, with the difference from others that it is run by millions of simple people who cultivate fruit and vegetables, breed livestock, tend to the earth, work and eat in the name of an interest that is greater than profit—more complex, hence less graspable.

The Enjoyment of Life

The local scale is an effective way of countering the impoverishment and the crisis of our present way of "living" on the earth. To achieve an "alternative" economy and a way of thinking more in line with the true essence of what we are, we have to address and solve problems of context, self-regulation, and space. We have to realize that we have long since gone too far, that our way of figuring things out isn't the same as nature's, that in all the processes we set into motion there are hidden connections, energy dispersions that have to be directed, understood as far as possible, and taken into account. As Georgescu-Roegen says, dispersed energies tell us how much we enjoy life.

The road we are traveling with local-economy systems will be a long one. Slowness is a value, but also a reminder that we can't have everything straightaway and that the things that count most are intentions, openness, memory, and care. It isn't the "New Word" because we recognize everyone's right to the benefit of the doubt, to the precautionary principle, and to make mistakes. It is, after all, in moments of error and crisis that man gives of his best. Which is why, faced with manifold crises—economic, financial, climatic, environmental, ethical—I propose a renaissance based on food, a new humanism founded on sentiment, not on calculation, to be enjoyed so that our activities aren't increasingly incompatible one with another and with our existence on the earth.

When they tell me that Terra Madre is pure make-believe, a pipe dream, I respond that it's not true, that it's humanity that's living in a world of make-believe. The communities have their feet on the ground and their hands in the soil; they are perfectly integrated in the context in which they live, and they draw enjoyment and benefits from that integration. I also believe that, among the men and women who populate the world today, it is increasingly difficult to find people who still know how to live life as it is meant to be lived, in peace.

There is, and there will be, no peace until we regenerate our relationship with the earth—that is, with food. By losing a healthy relationship with food, we have allowed the earth, once a kind and tender mother, to turn into a wicked stepmother. And, I have to acknowledge the fact, it had every right to do so. Let's take care of our local areas once more, of our neighbors, of our people. Let's start eating our food again without being eaten by it, let's look beyond the ends of our noses, let's see what's happening in the world, not only what the TV and the newspapers tell us. Let's regain possession of our senses to understand our surroundings. Let's regain control of reality. Let's do what food communities do; let's adopt a lifestyle that is practicable everywhere—the nice thing being that it costs us no wearisome effort. For all the energy we invest will be restored to us in the form of pleasure.

SEVEN

the future of terra madre

Terra Madre is a network of communities, hence of people; real people who work with food and for food; people who put food at the center of their lives so as not to lose or let others lose the "enjoyment of life"; people who don't want to allow themselves to be eaten by food, but want instead to regain control of their vital relationship with the earth and all the living beings that inhabit it.

Terra Madre is participatory democracy and real, local, natural economy. It puts people, not multinationals, into the countryside and onto the seas. It makes food first for consumption, not for sale. It celebrates the joys of life in all its aspects and consciously seeks to build a better world. Terra Madre knows that this is the way to respond to crises and uncertainty. It knows where humanity is going and has all the answers for this postmodern age, in which people are losing their way and feeling increasingly lonely.

Terra Madre is a political subject, its poetry and aesthetics a celebration of all humanity. Above all, Terra Madre is a project with long-term, maybe limitless goals. It is a series of short but decisive steps toward a new humanism, a renaissance, which, like the Renaissance, begins with beauty. By beauty I mean good, clean, and fair food, our villages and our landscapes, not to mention our relationship with nature, which generates beauty and teaches us how to enjoy it.

Terra Madre is a project with a boundless capacity to build and create. It looks to the future without uncertainty but with great faith in systems we have yet to fully comprehend; systems made up of hidden connections that have enabled us to exist on the earth until now. To build, Terra Madre has to be sustainable; albeit not without contradictions, it proceeds on its course, seeking to improve all the time.

Terra Madre has to be free from ties with politics and economics, with the systems of power that govern the world today; not so much for reasons of "protest" or snobbery as for its need to act as a creative workshop. Its principles are unshakable and shared by its members; this is why the network has to extend to all those who wish to avoid becoming mere cogs, interchangeable parts in the global industrial machinery, to those who decide to sustain the dream of a better humanity with every means possible, each according to his or her own possibilities. Terra Madre needs resources to organize its gatherings; it needs the time and money of people who believe in it and wish to take part. It is a priceless adventure, a project capable of regenerating and fueling itself. But it also needs support because, notwithstanding the great results already achieved, it is still only at an initial stage.

It isn't impossible to draw a picture of what Terra Madre might become in the future, but it should be added that, insofar as the network exists and operates in the most diverse places on Earth, it could assume any form.

One of the project's most recent initiatives was Terra Madre Day, held on December 10, 2009, to mark the twentieth anniversary of Slow Food International and its manifesto. On that day, all the world's food communities and all the Slow Food convivia (chapters), presidium projects, gardens, and people associated with the network organized major and minor events to celebrate. From now on the Terra Madre event will be staged every two years as always, but Terra Madre Day will be held every year as a moment of global reflection. On December 10, many different places around the world echoed to the voices, music, and songs of the food communities, all keen to share their joy with others. Terra Madre Day will be a new way to make us feel together even when we are far apart, sure in the knowledge that other people elsewhere are working like us for the cause of better, more sustainable, more democratic food. Last year the celebration sparked great enthusiasm and renewed passion, and left its mark on all the food communities. Who knows what all this will generate at the local level?

Terra Madre will have to become increasingly autonomous at the local level. It needs to attract patrons who, instead of only investing money to create beauty and works of art, will finance the beauty of food, agriculture, nature, sustainability, and happiness. With the "austere anarchy" I spoke of, respect for the spirit of the initiative and the creativity of everyone involved, it should be possible to achieve financial decentralization; to find the resources to enable delegates of food communities to come to Turin every two years; to organize initiatives locally; to network more and more; to carry out research to share and exchange knowledge; and to travel when necessary (the younger generation in particular).

In short, the future of Terra Madre will be shaped in the name of freedom of initiative and food and existential sovereignty. In a certain sense, Terra Madre has to free itself from itself, from the people who devised it and organize it. It must adhere to its own reality, which is made up of millions, even billions of other different realities, of human lives that deserve the same respect as nature. No one can be allowed take away their right to live happily.

Terra Madre will be increasingly local, hence increasingly sustainable, effective, and, above all—and this isn't a paradox—global, in defiance of those who label the advocates of the local economy as "anti-global." We, the people of Terra Madre, are the most "global" of all, because we are well aware that we are a living, active, creative part of that most wonderful of globes—our Mother Earth.

A LETTER FROM ENZO BIANCHI, PRIOR OF BOSE

My dear Carlin,

The package containing the proofs of your book was waiting for me on my return from Paris, where I had been to open a conference at the Collège des Bernardins to mark the 800th anniversary of the approval of the Franciscan order. I was reminded of St. Francis on a number of occasions in the course of my stay. On the sidewalk in Avenue Voltaire, for example, I read the words: "L'âge de l'or était l'âge où l'or ne régnait pas" (The golden age was the age in which gold didn't reign.) Then I saw a movie poster that said, "Every man dies, not every man really lives," as if to say that it's better to add life to days, than days to life!

With these echoes in my heart, here I am now browsing through your manuscript. As I read your opening speech at Terra Madre in Turin on October 23, 2008, in my mind's eye I can see again the captivated faces of the farmers, fishermen, nomadic shepherds, cooks, and young people from 153 countries in the world, our beloved world. I'm reminded of the visionary words of Helder Camara: "When fatigue impregnates the clothes of the humble, look around you and you'll see that the angels are picking up beads of sweat as if they were picking up diamonds."

Yes, we share the responsibility, entrusted to us by the Creator, of taking care of his creation. God's covenant is with humanity and every other living creature "for perpetual generations" (Genesis 9:12). "Dominion ... over every living thing that moveth the earth" (Genesis 1:28) means that we have to take care of the earth as God would. "And the Lord God took the man, and put him into the Garden of Eden to dress it and keep it" (Genesis 2:15) means serving the earth and the sea, protecting them against profanation and exploitation. This becomes a fundamental prayer: "Through heaven and earth and the sea, through wood and stone, through all creation visible and invisible I render obeisance and honor to the one Creator" (Leontius of Neapolis, *Third Treatise on Holy Images*), because "The God of matter . . . became matter and deigned to inhabit matter [and] worked out my salvation through matter" (John of Damascus, *In Defense of Icons*, 1:16).

We are grateful to you for stressing the gravity and urgency of the present situation: to proceed along the way of excess and waste, to the detriment of our neighbors and the vital capacities of the planet, is not only irresponsible and shameful; it is, above all, suicidal because it puts the survival of creation, of the earth we share, at risk. Yet, it's true, there is still hope left. Modestly and generously, Terra Madre may be a "little light, like a rushlight / To lead back to splendour," to quote Ezra Pound at the end of his *Cantos*. Provided we never ask for whom the bell tolls—because it tolls for us. And provided we understand that we would never achieve what is possible, if we didn't aspire to the impossible.

Terra Madre may help not to let faith and hope die, to convey a sense of windows opening on a May dawn. To open those windows and show young people the shadows of a starless evening—and never again the sun—would be an act unworthy of man and would extinguish life.

My wish to you, Friends of Terra Madre, is that you do not fear getting lost on account of your passion, because, in any case, you would lose less than if you lost your passion. And to you, Carlin, I wish you the awareness that what you can do is a drop in the ocean. But it is what gives meaning to your life. And much more.
Thanks!

        Enzo Bianchi, Prior of Bose
        Monastery of Bose, October 9, 2009
        Feast of St. Abraham, Friend of God
        Father of All Believers in One God

Chelsea Green Publishing is committed to preserving ancient forests and natural resources. We elected to print this title on 30-percent postconsumer recycled paper, processed chlorine-free. As a result, for this printing, we have saved:

**7 Trees (40' tall and 6-8" diameter)
2 Million BTUs of Total Energy
682 Pounds of Greenhouse Gases
3,282 Gallons of Wastewater
199 Pounds of Solid Waste**

Chelsea Green Publishing made this paper choice because we and our printer, Thomson-Shore, Inc., are members of the Green Press Initiative, a nonprofit program dedicated to supporting authors, publishers, and suppliers in their efforts to reduce their use of fiber obtained from endangered forests. For more information, visit: www.greenpressinitiative.org.

Environmental impact estimates were made using the Environmental Defense Paper Calculator. For more information visit: www.papercalculator.org.

the politics and practice of sustainable living
CHELSEA GREEN PUBLISHING

Chelsea Green Publishing sees books as tools for effecting cultural change and seeks to empower citizens to participate in reclaiming our global commons and become its impassioned stewards. If you enjoyed *Terra Madre*, please consider these other great books related to sustainable agriculture.

INQUIRIES INTO THE NATURE OF SLOW MONEY
Investing as if Food, Farms, and Fertility Mattered
WOODY TASCH
Foreword by CARLO PETRINI
9781603580069
Hardcover • $21.95

SLOW FOOD
Collected Thoughts on Taste, Tradition, and the Honest Pleasure of Food
Edited by CARLO PETRINI and BEN WATSON
9781931498012
Paperback • $24.95

CHEESEMONGER
A Life on the Wedge
GORDON EDGAR
9781603582377
Paperback • $17.95

LIBATION, A BITTER ALCHEMY
DEIRDRE HEEKIN
9781603580861
Hardcover • $25.00

For more information or to request a catalog, visit **www.chelseagreen.com** or call toll-free **(800) 639-4099**.

the politics and practice of sustainable living
CHELSEA GREEN PUBLISHING

WILD FERMENTATION
*The Flavor, Nutrition,
and Craft of Live-Culture Foods*
SANDOR KATZ
Foreword by SALLY FALLON
9781931498234
Paperback • $25.00

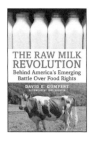

THE RAW MILK REVOLUTION
*Behind America's Emerging
Battle Over Food Rights*
DAVID E. GUMPERT
Foreword by JOEL SALATIN
9781603582193
Paperback • $19.95

GAIA'S GARDEN
A Guide to Home-Scale Permaculture
TOBY HEMENWAY
9781603580298
Paperback • $29.95

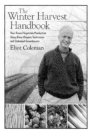

THE WINTER HARVEST HANDBOOK
*Year-Round Vegetable Production Using Deep
Organic Techniques and Unheated Greenhouses*
ELIOT COLEMAN
9781603580816
Paperback • $29.95

For more information or to request a catalog,
visit **www.chelseagreen.com** or
call toll-free **(800) 639-4099**.